Success guides

Leckie×Leckie

Scotland's leading educational publishers

Standard Grade
Chemistry

Archie Gibb × Emma Poole

Contents

Contents

Making electricity: electrochemical cells

Metals and corrosion

Plastics and synthetic fibres

Making Fertilisers

Carbohydrates to alcohol

Calculations

Elements, compounds and mixtures

Elements

Everything in the world is made from just over 100 elements. Every element is made up of very small particles called **atoms.** An **element** is a substance made up of one type of atom.

The names, symbols and atomic numbers of the different elements are given in the Periodic Table on page 8 of your Data Booklet:

- Magnesium is an element and it has the symbol, Mg. All magnesium atoms are the same.
- Oxygen is another element. It has the symbol, O. All oxygen atoms are the same but are different from magnesium atoms.

Top Tip
Remember that the Periodic Table in your Data Booklet has the names and symbols of the elements.

Compounds

Compounds are formed when elements react together. A **compound** contains two or more elements joined together. For example, magnesium oxide is a compound formed when the elements magnesium and oxygen react together. You may have seen magnesium burning in air. When this happens the magnesium is reacting with the oxygen in the air.

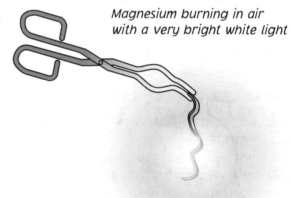

Magnesium burning in air with a very bright white light

The **word equation** that describes this reaction is:

magnesium + oxygen → magnesium oxide
      elements   compound

Compounds with names ending in **-ide** contain the two elements indicated in the name. Magnesium ox**ide**, MgO, contains the two elements magnesium and oxygen.

Compounds with names ending in **-ate** or **-ite** also contain the element oxygen. Sodium carbon**ate**, Na_2CO_3, contains the elements sodium, carbon **and oxygen**.

Top Tip
Remember: if the name of the compound ends in -ite or -ate, then the compound also contains oxygen.

Mixtures

When two substances are mixed together but don't actually react with each other we say that a **mixture** has been formed. Common mixtures include air and crude oil. Air is a mixture of nitrogen, oxygen and other gases. Crude oil is a mixture of different hydrocarbons.

Solutions

When a **solute** dissolves in a **solvent** a **solution** is formed. A solution is a mixture, because the solute has not reacted with the solvent when it dissolves.

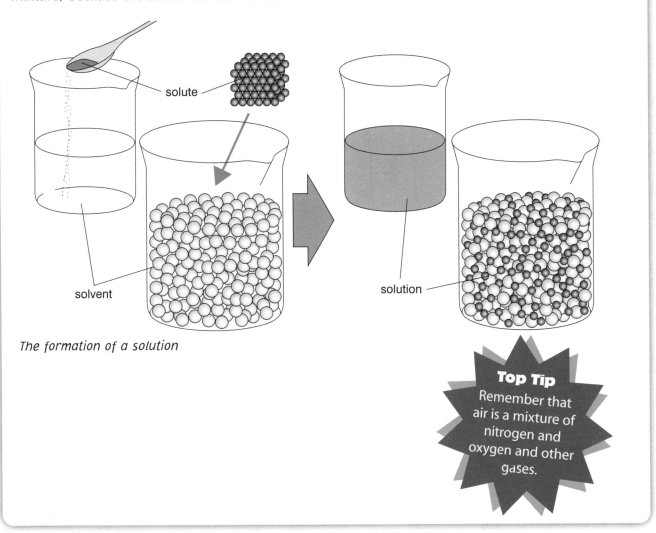

solute

solvent

solution

The formation of a solution

Top Tip
Remember that air is a mixture of nitrogen and oxygen and other gases.

Quick Test

1. Write down the symbols for the following elements:
   magnesium, chlorine, copper, iron, silver

2. Which elements are present in:
   copper oxide, zinc carbonate, sodium sulphite, potassium nitrate, sodium chloride, sodium chlorate?

3. What is the difference between a compound and a mixture?

Answers 1 Mg Cl Cu Fe Ag **2** copper and oxygen; zinc, carbon and oxygen; sodium, sulphur and oxygen; potassium, nitrogen and oxygen; sodium and chlorine; sodium, chlorine and oxygen. **3** In a compound the elements are joined or combined together, whereas in a mixture they are not.

Identifying chemical reactions

When a chemical reaction takes place a new substance is formed. The substances at the start of the reaction are known as the **reactants** and the new substances are known as the **products**.

What to look for in a reaction

Three things that indicate a chemical reaction has taken place include:

- a colour change

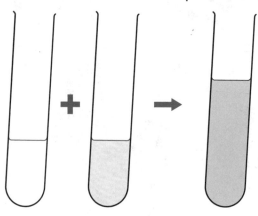

- a precipitate forming

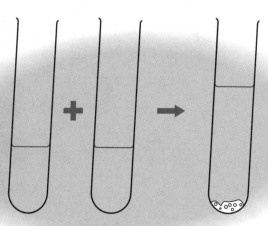

Top Tip
Remember that a precipitate is the solid formed when two liquids are mixed together.

- a gas being given off

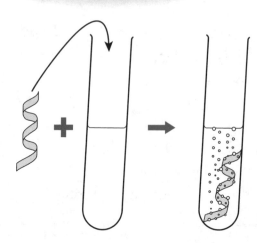

Exothermic and endothermic reactions

Sometimes you cannot see any change when a chemical reaction takes place but there may be an energy change such as a change in temperature.

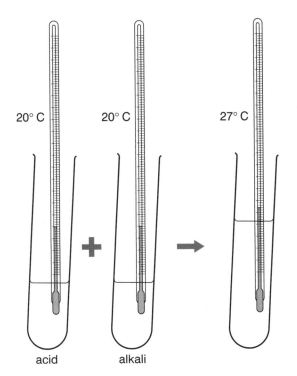

20° C   20° C   27° C

acid   alkali

an acid reacting with an alkali to produce an increase in temperature

Top Tip
You should recognise that a chemical reaction is taking place if you see
*bubbles
*a colour change
*a precipitate forming
*that heat has been given out or taken in.

A reaction in which heat is given out is **exothermic** (heat given out). If the temperature had dropped the reaction would be **endothermic** (heat taken in)

Quick Test

1. Which of the changes below are chemical reactions?
   a) melting ice
   b) breaking glass
   c) frying an egg
   d) hydrogen produced when magnesium is added to acid
   e) adding water to orange juice
   f) iron rusting
   g) separating alcohol from water by distilling
   h) filtering sand from water.

2. What does exothermic mean?

Answers 1. c, d and f. 2. Heat energy is given out.

Rates of reaction

Rates of reaction can be
← slow or fast →
rusting   explosions

Analysing rates of reaction

The rate of a chemical reaction can be measured by:

- how fast the **products** are being made
- how fast the **reactants** are being used up.

The graph (at right) shows the amount of product made in three experiments.

The graph is **steepest** at the start of the reaction for all three experiments; it then starts to **level out** as the reactant particles get used up.

When the graph becomes level, the reaction has finished. The graph shows that **Experiment 2** is faster than **Experiment 1**.

This could have been due to Experiment 2 having a higher temperature, a greater concentration of reactant particles (or pressure for gases), a greater surface area or a catalyst being added to it.

Experiments 1 and 2 make the **same amount** of the product so they both had the same amounts of reactants at the start of the reaction.

In Experiment 3 only **half** as much product was made.

This shows that there were less reactants at the start of Experiment 3.

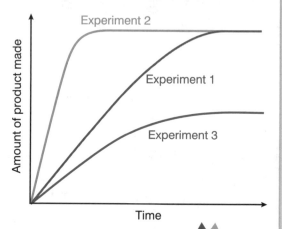

Top Tip
When analysing graphs remember:
• the steeper the slope the faster is the reaction
• the reaction is over when the graph levels out.

Increasing the surface area

The greater the **surface area**, the more chance of collisions occurring, so the faster the rate of reaction.

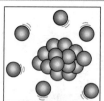

small surface area (larger pieces)

large surface area (smaller pieces)

Top Tip
Remember: small pieces have a larger overall surface area.

Temperature

If the temperature is increased, particles move quicker. Increasing the temperature increases the **rate** of **reaction** because the particles collide more often and with more energy so there are more successful collisions, and the rate increases.

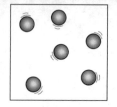

*a **low temperature** slows down reaction rate.*

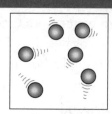

*a **high temperature** speeds up reaction rate.*

Increasing the concentration of dissolved reactants

The **greater** the concentration, the **more** reactant particles there are in the solution. There will be more collisions and so the reaction rate is increased.

low concentration of particles

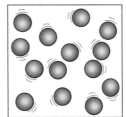

high concentration of particles

Concentrated acid will react more quickly than dilute acid.

Adding a catalyst

- A catalyst increases the rate of reaction, but is not itself used up in the reaction.
- Catalysts are specific to certain reactions.
- An everyday example of a catalyst is platinum and other transition metals inside the catalytic converters in car exhausts.

Quick Test

1. What happens to the rate of a reaction if the temperature is decreased?
2. What happens to the rate of a reaction if the concentration of the dissolved reactants is increased?
3. What happens to the rate of a reaction if the surface area of a reactant is increased?
4. What does a catalyst do to the rate of reaction?
5. Why can catalysts be reused?
6. How can the rate of a chemical reaction be measured?
7. Give four ways in which you could increase the rate of a chemical reaction.
8. Give three ways in which you could decrease the rate of a chemical reaction.

Answers 1. Decreases **2.** Increases **3.** Increases **4.** A catalyst increases the rate of reaction **5.** Not used up during the reaction **6.** How fast reactant used up/product is made **7.** Increase temperature; increase concentration; increase surface area; add a catalyst **8.** Decrease temperature; decrease concentration; decrease surface area

Classifying elements

Solids, liquids or gases?

Most chemical elements are **solid** at room temperature. Some elements exist as **gases** but only two elements, mercury (Hg) and bromine (Br) are **liquid** at room temperature. The state at room temperature (approximately 25 °C) can be determined by looking at the **melting and boiling points** of the elements on page 3 of the Data Booklet.

- If 25 °C is above the melting point of the element, then the element is a solid at room temperature.
- If 25 °C is between the melting and boiling points of the element, then the element is a liquid at room temperature.
- If 25 °C is above the boiling point of the element, then the element is a gas at room temperature.

Top Tip
Make sure you know where you can find information in your Data Booklet. It is a very useful examination resource.

Metals or non-metals?

Three-quarters of the elements in the Periodic Table are **metals.** Elements shown at the left-hand side of the dark zig-zag line on the Periodic Table are metals, except for hydrogen, element number 1, which is a non-metal. This is seen on the Periodic Table opposite. This is the same as the Periodic Table on page 8 of the Data Booklet.

Naturally occurring or made by scientists?

Most elements are found naturally either as the element or in compounds. The elements after uranium (number 92) have been made by scientists and do not occur naturally. These can be seen on the Periodic Table opposite.

Quick Test

1. Name and write the symbol for the only metal which is not solid at room temperature.
2. Which other element is liquid at room temperature?
3. Which three elements in Group 4 are metals?
4. Which element in Group 5 is a gas at room temperature?
5. In which group are all the elements gases?
6. Name two elements that are not naturally occurring and have been made by scientists.
7. Name an element in Group 7 that is solid at room temperature.
8. Which two elements in Group 7 are gases at room temperature?
9. Use the Periodic Table to find out what you can about sodium and xenon.

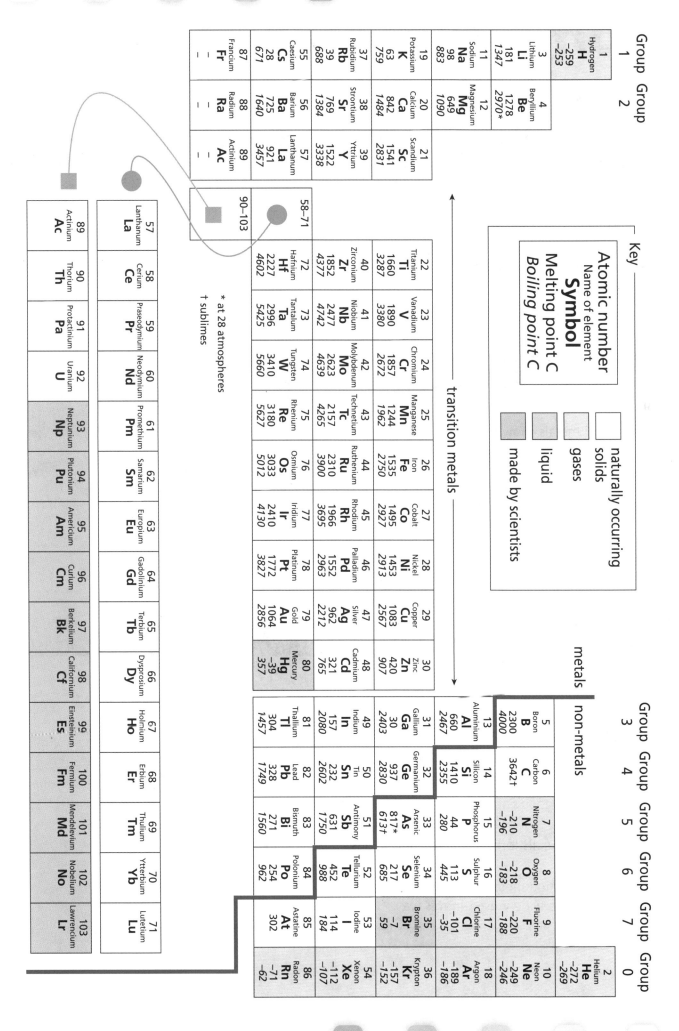

Some families in the Periodic Table

The Periodic Table

- In the modern Periodic Table the elements are arranged in order of increasing atomic number.
- The elements are placed in rows so that elements with similar properties are in the same column.
- These vertical columns are called **groups**.
- All the members of a group share similar properties.
- The horizontal rows are called **periods**.

Top Tip
Elements in the same group have similar chemical properties.

Group 1: The alkali metals

Top Tip
The alkali metals are very reactive metals and must be stored under oil.

Examples: **lithium**, **sodium**, **potassium**

The alkali metals show these properties:

- They have **lower densities** than water and are quite soft compared to other metals.
- They have **low melting points** compared with other metals.
- They **react vigorously** with water, releasing **hydrogen gas** and forming **alkaline solutions**.

The Group 1 metals react with non-metals to form white, ionic compounds which **dissolve** to form colourless solutions.

The transition metals

Examples: **iron**, **nickel** and **copper**

- Transition metals are **hard** and **strong**.
- They are much **less reactive** than the metals in Group 1 and do not react quickly with oxygen or water.
- Transition metals are widely used.
- Iron is often used as a **structural material**.
- Copper is a good conductor of both heat and electricity, and it is often used for electrical cables.
- Transition metals form coloured compounds which can be used in pottery glazes.
- Many transition metals and their compounds can act as catalysts. Iron and platinum are widely used in this way.

Group 7: The halogens

Fluorine

- Fluorine is a **very poisonous**, pale yellow gas.

Chlorine

- Chlorine is a **poisonous**, pale green gas.
- Chlorine is used in water purification and bleaching.

Bromine

- Bromine is a **poisonous**, dense, brown **liquid**.

Iodine

- Iodine is a dark grey, crystalline **solid** or a purple vapour.
- Iodine solution is used as an antiseptic.

Down the group each element becomes less reactive.

The halogens have coloured vapours.

Down the group the colour of the vapour gets darker.

- Melting and boiling points increase down the group (the first two are gases, the next – bromine – is a liquid, and iodine is a solid).
- All the halogens are poisonous and should only be used in a fume cupboard.
- Halogens react with metals to form compounds.
- Halogens can also **react with non-metals** to form **covalently bonded compounds**.

Top Tip
The halogens are the most reactive non-metal elements.

Quick Test

1. What are the vertical columns in the Periodic Table called?
2. Give three examples of alkali metals.
3. Why do alkali metals float on water?
4. Why are alkali metals stored under oil?
5. Which gas is released when alkali metals react with water?
6. Name three transition metals.
7. Give two uses of transition metals.
8. What is the name given to the Group 7 elements?
9. Which is the most reactive element in Group 7?
10. What is chlorine used for?

Group 0: the noble gases

Chemical properties of the noble gases

- The noble gases are the elements helium, neon, argon, krypton, xenon and radon.
- They are very unreactive gases.
- They are sometimes called 'inert' because they are so unreactive.
- Argon makes up about 1 per cent of dry air.

Trends in the noble gases

As you go down the group:

- density increases;
- boiling point increases.

The noble gases are colourless, monatomic gases (they exist as individual atoms rather than as diatomic molecules as other gases do).

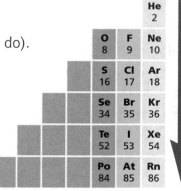

5	6	7	0
			He 2
	O 8	F 9	Ne 10
	S 16	Cl 17	Ar 18
	Se 34	Br 35	Kr 36
	Te 52	I 53	Xe 54
	Po 84	At 85	Rn 86

Physical properties of the noble gases

- All the members of Group 0 are **gases at room temperature**.
- If they are **cooled down** they can become **liquids** and eventually **solids**.
- They are poor conductors of both electricity and heat.

Uses of the noble gases

Helium

- Helium is used in **balloons** and in **airships**, because it is **less dense** than air (and **not flammable** like hydrogen).

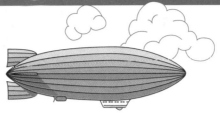

Argon

- Argon is used in light bulbs (**filament lamps**).
- Surrounding the hot filament with inert argon stops it from burning away.

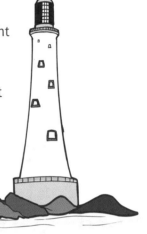

Neon

- Neon is used in electrical discharge tubes in advertising signs.

Krypton

- Krypton is used in **lasers**.

Quick Test

1. Name and give the symbols for the first three noble gases.
2. How much argon will be present in 1 litre of dry air?
3. What does 'monatomic' mean?
4. What is helium used for?
5. What is argon used for?
6. What is neon used for?
7. Which noble gas has the lowest boiling point?
8. Which noble gas has the greatest density?

Answers 1. Helium (He), Neon (Ne) and Argon (Ar). **2.** 10 cm³. **3.** Existing as individual atoms. **4.** Balloons and airships. **5.** It is the gas inside light bulbs. **6.** Advertising signs. **7.** Helium. **8.** Radon.

Atomic structure

Atoms

Every element is made up of very small particles called **atoms.** Atoms have a very small positively charged **nucleus** with negatively charged **electrons** moving around outside the nucleus. An atom is neutral because the positive charge of the nucleus is equal to the sum of the negatively charged electrons outside the nucleus.

Top Tip
Make sure you are familiar with the charge, mass and location of the three types of particles.

The structure of an atom

An atom has a nucleus surrounded by shells of electrons.

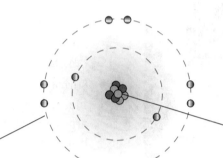

The **nucleus** is found at the centre of the **atom** and contains **neutrons** and **protons**.

The **electrons** are found in shells around the **nucleus**.

Structure of the atom

- **Protons** have a **positive charge** and a **mass of 1.**
- **Neutrons** have **no charge** and also have a **mass of 1.**
- **Electrons** have a **negative charge** and a **negligible mass** (which we take as 0).

Top Tip
Number of neutrons = mass number – atomic number

Information about protons, neutrons and electrons is summarised in the table below:

Particle	Mass	Charge	Location
Proton	1	+1	In the nucleus
Neutron	1	0	In the nucleus
Electrons	0	−1	Outside the nucleus

In all neutral atoms there is no overall charge, so the number of protons is equal to the number of electrons.

$^{23}_{11}$**Na** The **mass number** is the number of protons added to the number of neutrons.

The **atomic number** is the number of protons.

- Sodium has an atomic number of 11, so it has 11 protons.
- The sodium atom has no overall charge so the number of electrons must be the same as the number of protons. Sodium therefore has 11 electrons.
- The number of neutrons is given by the mass number minus the atomic number.
- Sodium has 23 − 11 = 12 neutrons.

Isotopes

Isotopes of an element have the **same** number of protons but a **different** number of neutrons. So isotopes have the same **atomic number** but a different **mass number**.

- Chlorine has 2 common isotopes:

$^{35}_{17}\text{Cl}$:
- 17 protons
- 17 electrons
- **18 neutrons**

$^{37}_{17}\text{Cl}$:
- 17 protons
- 17 electrons
- **20 neutrons**

- The isotopes will react chemically in the same way because they have identical numbers of electrons.
- Most elements exist as a mixture of isotopes.

The atomic structure of elements 1–20

CREDIT

Symbol	Elements	Protons	Electrons	Neutrons
$^1_1$ H	Hydrogen	1	1	0
$^4_2$ He	Helium	2	2	2
$^7_3$ Li	Lithium	3	3	4
$^9_4$ Be	Beryllium	4	4	5
$^{11}_5$ B	Boron	5	5	6
$^{12}_6$ C	Carbon	6	6	6
$^{14}_7$ N	Nitrogen	7	7	7
$^{16}_8$ O	Oxygen	8	8	8
$^{19}_9$ F	Fluorine	9	9	10
$^{20}_{10}$ Ne	Neon	10	10	10

Symbol	Elements	Protons	Electrons	Neutrons
$^{23}_{11}$ Na	Sodium	11	11	12
$^{24}_{12}$ Mg	Magnesium	12	12	12
$^{27}_{13}$ Al	Aluminium	13	13	14
$^{28}_{14}$ Si	Silicon	14	14	14
$^{31}_{15}$ P	Phosphorus	15	15	16
$^{32}_{16}$ S	Sulphur	16	16	16
$^{35}_{17}$ Cl	Chlorine	17	17	18
$^{40}_{18}$ Ar	Argon	18	18	22
$^{39}_{19}$ K	Potassium	19	19	20
$^{40}_{20}$ Ca	Calcium	20	20	20

Quick Test

1. What does the nucleus contain?

2. What are found in shells around the nucleus?

3. What is the charge and mass of a proton?

4. What is the charge and mass of an electron?

5. What is the charge and mass of a neutron?

6. What is the mass number of an atom?

7. What is the atomic number of an atom?

8. What is the same about the atoms of two isotopes of an element?

9. What is different about the atoms of two isotopes of an element?

10. Why do isotopes of an element react in the same way?

11. Calculate the number of protons, neutrons and electrons in $^{63}_{29}$Cu.

Answers 1. Protons and neutrons 2. Electrons 3. Charge +1, mass 1 4. Charge −1, mass negligible 5. No charge, mass 1 6. Number of protons + number of neutrons 7. Number of protons. 8. Atomic number/number of protons or electrons 9. Mass number/number of neutrons 10. They have the same number of electrons. 11. 29 protons, 34 neutrons and 29 electrons.

Electron arrangement

Electron arrangement

- Electrons are arranged in energy levels or shells. `CREDIT`
- The first shell may contain up to **2** electrons.
- The second shell contains up to **8** electrons.
- The third shell can hold up to **18** electrons.
- The number of electrons in the outer shell indicates the group that the element belongs to. So, the electron arrangement shows how the atom will react chemically.
- The maximum number of electrons in the **outer shell** is **8**.

Top Tip
The electron arrangements are given on page 1 of the Data Booklet.

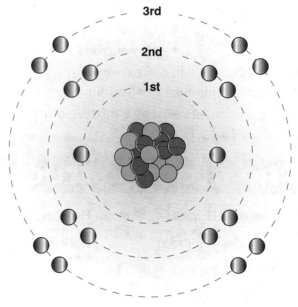

A model of electron shells

Lithium
- Number of protons = 3.
- Number of electrons = 3.
- The electron arrangement is 2, 1.

Magnesium
- Number of protons = 12.
- Number of electrons = 12.
- The electron arrangement is 2, 8, 2.

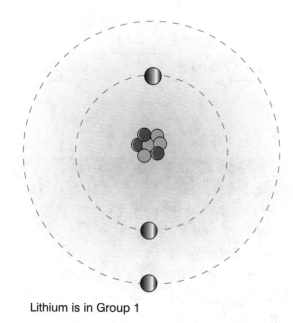

Lithium is in Group 1

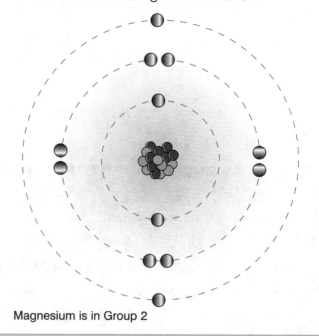

Magnesium is in Group 2

The electron arrangement of elements 1–20

Symbol	Element	Electrons	Electron arrangement	Group
H	Hydrogen	1	1	1
He	Helium	2	2	0
Li	Lithium	3	2, 1	1
Be	Beryllium	4	2, 2	2
B	Boron	5	2, 3	3
C	Carbon	6	2, 4	4
N	Nitrogen	7	2, 5	5
O	Oxygen	8	2, 6	6
F	Fluorine	9	2, 7	7
Ne	Neon	10	2, 8	0
Na	Sodium	11	2, 8, 1	1
Mg	Magnesium	12	2, 8, 2	2
Al	Aluminium	13	2, 8, 3	3
Si	Silicon	14	2, 8, 4	4
P	Phosphorus	15	2, 8, 5	5
S	Sulphur	16	2, 8, 6	6
Cl	Chlorine	17	2, 8, 7	7
Ar	Argon	18	2, 8, 8	0
K	Potassium	19	2, 8, 8, 1	1
Ca	Calcium	20	2, 8, 8, 2	2

Top Tip
The number of outer electrons determines the reactivity of the element.

Top Tip
Note that elements with the same number of outer electrons have similar chemical properties. For example, the alkali metals are very reactive as all their atoms have one outer electron.

Quick Test

1. Where are electrons found?
2. How many electrons can the first shell hold?
3. What is the maximum number of electrons in the outer shell of an atom?
4. If an atom has seven protons, how many electrons will it have?
5. Write down the electron arrangement for sodium.
6. How many outer electrons are there in atoms of the alkali metals?
7. How many outer electrons are there in atoms of the halogens?

Answers 1. Around the nucleus in shells or energy levels. 2. Up to two. 3. Eight. 4. Seven. 5. 2, 8, 1. 6. One. 7. Seven.

Covalent bonding

How it works

Atoms can be held together by **bonds. Covalent bonding** occurs between **non-metals.** The atoms **share** electrons in the bond. Covalent bonding allows both atoms to have a stable, full **outer shell.**

A covalent bond is when the two positive nuclei are held together by their common attraction for the shared pair of electrons.

> **Top Tip**
> Remember that a covalent bond is a pair of shared electrons.

In the following diagrams, the outer electrons from one atom are represented by dots and the outer electrons from the other atom are represented by crosses.

Hydrogen, H_2

Both hydrogen atoms have only one electron, but by forming a single covalent bond, both can have a full outer shell.

This can also be shown as **H—H**

Hydrogen chloride, HCl

The hydrogen and the chlorine atoms both need **one** more electron. They form a single covalent bond so both have a full outer shell.

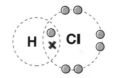

This can also be shown as

H—Cl

Molecules such as H_2 and HCl are **diatomic** (there are two atoms in a molecule). **Elements** which exist as **diatomic molecules** are H_2, N_2, O_2, F_2, Cl_2, Br_2 and I_2.

Methane, CH_4

The carbon has four outer electrons so needs **four** more for a full outer shell. The carbon forms four single covalent bonds to the hydrogen atoms, so all the atoms now have a full outer shell of electrons.

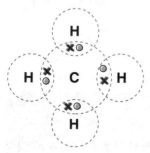

This can also be shown as

Ammonia, NH_3

The nitrogen atom has five outer electrons so needs three more. Nitrogen forms three single covalent bonds to hydrogen atoms.

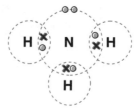

This can also be shown as

> **Top Tip**
> You need to be able to draw the shapes of water and methane molecules.

Water, H₂O

The oxygen has six outer electrons so needs two more. The oxygen forms two single covalent bonds with the two hydrogen atoms to give it a full outer shell.

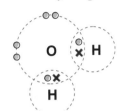

This can also be shown as

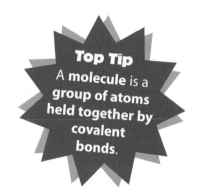

Oxygen, O₂

Both oxygen atoms have six outer electrons so both need two more. The oxygen atoms form one double covalent bond so that both have a full outer shell.

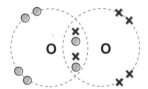

This can also be shown as

O = O

Carbon dioxide, CO₂

The carbon has four outer electrons, so needs four more. It forms double covalent bonds with two oxygen atoms, so that all the atoms now have a full outer shell of electrons.

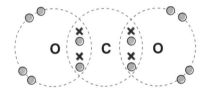

This can also be shown as

O = C = O

Quick Test

1. Which elements are involved in covalent bonding?

2. How many outer electrons do elements in Group 6 have?

3. What is meant by a molecule?

4. Draw a dot and cross diagram to show hydrogen bonding with chlorine.

5. Draw a dot and cross diagram to show the bonding in methane.

6. What is the chemical formula for methane?

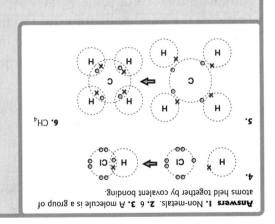

Answers 1. Non-metals. **2.** 6 **3.** A molecule is a group of atoms held together by covalent bonding.

6. CH₄

Chemical formulae

What the chemical formula means

The **chemical formula** shows the number of atoms of each element in a molecule. For example, water has the chemical formula H_2O. This means that each water molecule contains two hydrogen atoms and one oxygen atom joined together.

Chemical formulae you must know

There are some formulae you should get to know. Some of these are shown in the table below.

You should also try to remember the elements that exist as **diatomic elements**.

Top Tip
Chemical formulae are very important. Make sure you can write chemical formulae correctly.

Name	Formula
Water	H_2O
Carbon dioxide	CO_2
Hydrochloric acid	HCl
Sulphuric acid	H_2SO_4
Nitric acid	HNO_3
Sodium hydroxide	$NaOH$
Hydrogen	H_2
Nitrogen	N_2
Oxygen	O_2
Fluorine	F_2
Chlorine	Cl_2
Bromine	Br_2
Iodine	I_2

Diatomic elements

Formulae from names of compounds

Sometimes the name of the compound gives information about the formula of that compound. Names of these compounds have prefixes that give the number of atoms of certain elements in each molecule. Prefixes and their meanings are given in the table on the right.

- Carbon monoxide contains one carbon atom joined to one oxygen atom, so it has the formula CO.
- Carbon dioxide contains one carbon atom joined to two oxygen atoms, so it has the formula CO_2.
- Phosphorus pentachloride has the formula PCl_5.
- Dinitrogen tetroxide has the formula N_2O_4.

Prefix	Meaning
mono-	one
di-	two
tri-	three
tetra-	four
penta-	five

Valency

Valency is the combining power of an element. It depends on the number of electrons available for bonding in the outer shell of atoms of that element. Elements in the same group of the Periodic Table will have the same valency as they have the same number of outer electrons.

Valencies of the elements in the different groups in the Periodic Table are shown in the table below. Note that the noble gases have valency of 0. They have no combining power as they are so stable that they do not usually combine with other elements.

Group number	1	2	3	4	5	6	7	0
Valency	1	2	3	4	3	2	1	0

Using the Periodic Table in your Data Booklet will help you find the valencies of different elements.

The valency is really the number of unpaired electrons in the outer shell of the atoms of that element. Only unpaired electrons are usually available for bonding.

Using your knowledge of the Periodic Table you should be able to work out the valencies of different elements.

- Sodium is in Group 1 and so Na will have valency = 1.
- Carbon is in Group 4, and so C has valency = 4.
- Oxygen is in Group 6 and so O has valency = 2
- The halogens are in Group 7 and so all the halogens have valency = 1.

Quick Test

1. Write down the formulae for the following:
   a) carbon dioxide
   b) hydrochloric acid
   c) fluorine
   d) hydrogen
   e) nitric acid
   f) sulphur trioxide
   g) carbon tetrachloride
   h) nitrogen monoxide
   i) sulphuric acid
   j) sodium hydroxide

2. Write down the valency for
   a) potassium
   b) nitrogen
   c) bromine
   d) argon
   e) calcium
   f) aluminium
   g) sulphur
   h) silicon
   i) phosphorus

Answers 1. a) CO_2 **b)** HCl **c)** F_2 **d)** H_2 **e)** HNO_3 **f)** SO_3 **g)** CCl_4 **h)** NO **i)** H_2SO_4 **j)** NaOH **2. a)** 1 **b)** 3 **c)** 1 **d)** 0 **e)** 2 **f)** 3 **g)** 2 **h)** 4 **i)** 3

Chemical formulae using valency

Using valencies to work out the chemical formula

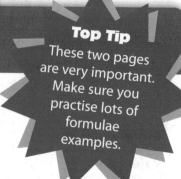

Top Tip

These two pages are very important. Make sure you practise lots of formulae examples.

The chemical formula of a compound is obtained by swapping over the valencies of the two elements.

To work out the formula of sodium sulphide:

- Sodium is in Group 1 and so has valency = 1
- Sulphur is in Group 6 and so has valency = 2
- We write this as $\overset{1}{Na}\overset{2}{S}$ and when the valencies are swapped over we get the formula Na_2S_1
- The '1' is not usually written, and so the formula would be correctly written as Na_2S

To work out the formula for magnesium chloride:

- Magnesium is in Group 2 and so has valency = 2
- Chlorine is in Group 7 and so has valency = 1
- Writing this as $\overset{2}{Mg}\overset{1}{Cl}$ and swapping over the valencies gives the formula $MgCl_2$

If both elements have the same valencies, then the valencies cancel out and the formula contains one atom of each element. An example of this is magnesium oxide in which both magnesium and oxygen have valency = 2 and so the formula is written as MgO

Formulae of compounds containing selected groups of atoms

These involve compounds containing larger groups of atoms and include formulae of compounds containing sulphate, nitrate and hydroxide ions. Formulae for these ions are given on page 4 of the Data Booklet. The size of the charge on the ion given in the Data Booklet gives the valency. For example, if the ion has a charge of 1+ or 1−, then it has valency = 1. If the charge is 2−, then the valency is 2. If the charge is 3−, then the valency is 3. The table right gives some examples.

Consider the sulphate group of atoms. It has formula SO_4 and valency = 2

- Working out the formula for sodium sulphate, sodium has valency = 1
  and sulphate has valency = 2
  $\overset{1}{Na}\overset{2}{SO_4}$ swapping over the valencies gives the formula Na_2SO_4

Name	Formula from the Data Booklet	Valency
Ammonium	NH_4^+	1
Hydroxide	OH^-	1
Nitrate	NO_3^-	1
Carbonate	CO_3^{2-}	2
Sulphate	SO_4^{3-}	2
Sulphite	SO_3^{2-}	2
Phosphate	PO_4^{3-}	3

- Calcium sulphate will be CaSO$_4$ since both calcium and sulphate have the same valency of 2 and so the valencies cancel out.
- Ammonium and nitrate both have valency of 1 and so the formula for ammonium nitrate is NH$_4$NO$_3$.

Elements which can have different valencies

CREDIT

Some elements, particularly the **transition metals**, do not always have the same valency in their different compounds. An example is copper which usually has valency of 2, but in some compounds it has valency of 1. The valency of these elements is usually given in roman numerals inside brackets.

- In copper(I) oxide the valency of copper is 1 and since oxygen has valency of 1, the formula is Cu$_2$O.
- The formula of copper(II) oxide is CuO.

Formulae containing brackets

CREDIT

Sometimes brackets are needed in formulae. An example is calcium nitrate.
- Calcium is in Group 2 and has valency of 2.
- Nitrate is NO$_3$ and has valency of 1.
- When the valencies are swapped over there will be one calcium in the formula and two nitrates. The formula must be written with nitrate inside brackets as Ca(NO$_3$)$_2$.

Writing the formula in this way shows that it contains one calcium, two nitrogen and six oxygen atoms. The '2' outside the brackets shows that everything inside the brackets is multiplied by 2. Other examples include calcium hydroxide, Ca(OH)$_2$ and ammonium sulphate, (NH$_4$)$_2$SO$_4$.

Quick Test

1. Write the chemical formula for the following:
   a) sodium oxide
   b) magnesium sulphide
   c) calcium chloride
   d) aluminium oxide
   e) magnesium nitride.

2. Write the chemical formula for the following:
   a) sodium carbonate
   b) magnesium sulphate
   c) potassium nitrate

3. Write the chemical formula for the following:
   a) copper(II) chloride
   b) nickel(II) oxide
   c) iron(III) chloride

4. Write the chemical formula for the following:
   a) iron(III) hydroxide
   b) magnesium nitrate
   c) ammonium sulphate
   d) ammonium phosphate
   e) ammonium carbonate

Answers 1. (a) Na$_2$O **b)** MgS **c)** CaCl$_2$ **d)** Al$_2$O$_3$ **e)** Mg$_3$N$_2$ **2. a)** Na$_2$CO$_3$ **b)** MgSO$_4$ **c)** KNO$_3$ **3. a)** CuCl$_2$ **b)** NiO **c)** FeCl$_3$ **4. a)** Fe(OH)$_3$ **b)** Mg(NO$_3$)$_2$ **c)** (NH$_4$)$_2$SO$_4$ **d)** (NH$_4$)$_3$PO$_4$ **e)** (NH$_4$)$_2$CO$_3$

Chemical equations

- Formula equations show the number of atoms of reactants and products.
- Since atoms cannot be created nor destroyed, there must be the same number of atoms of each element on each side of the equation. For example:

magnesium + oxygen → magnesium oxide

$2Mg$ + O_2 → $2MgO$

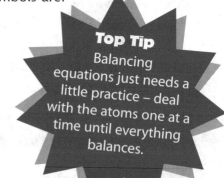

State symbols

State symbols can be added to an equation to give **extra information**.
They show what state the reactants and products are in. The symbols are:

- (s) for solid
- (l) for liquid
- (g) for gas
- (aq) for aqueous, or dissolved in water

Example

magnesium + oxygen → magnesium oxide

$2Mg(s)$ + $O_2(g)$ → $2MgO(s)$

Top Tip
Balancing equations just needs a little practice – deal with the atoms one at a time until everything balances.

Word equations

What happens in a chemical reaction can be summarised in a word equation.
The **reactants** are written on the left hand side of the equation and the
products are written on the right hand side of the equation. This is shown as:

reactants → products

where the → means 'changes into'.

For example, when magnesium burns in air the magnesium is reacting with
oxygen in the air making magnesium oxide. The word equation is:

magnesium + oxygen → magnesium oxide

Formula equations

When **hydrogen** burns in **oxygen**, **water** is made. The word equation is:

hydrogen + oxygen → water

Putting in the correct formulae gives the formula equation:

H_2 + O_2 → H_2O

The **formulae** are correct, but the equation is **not** balanced because there
are different numbers of atoms on each side of the equation.
The formulae **cannot** be changed, but the numbers in front of the formulae **can**
be changed.

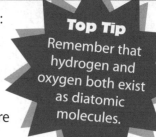

Top Tip
Remember that hydrogen and oxygen both exist as diatomic molecules.

How to balance an equation

$$H_2 + O_2 \rightarrow H_2O$$

Looking at the equation for hydrogen burning in oxygen we can see that there are **two** oxygen atoms on the left-hand side but only **one** on the right-hand side.

So a **2** is placed **in front** of the H_2O:

$$H_2 + O_2 \rightarrow 2H_2O$$

Now the oxygen atoms are balanced, but while there are two hydrogen atoms on the left-hand side there are now four hydrogen atoms on the right-hand side.

So a 2 is placed in front of the H_2:

$$2H_2 + O_2 \rightarrow 2H_2O$$

The equation is then balanced.

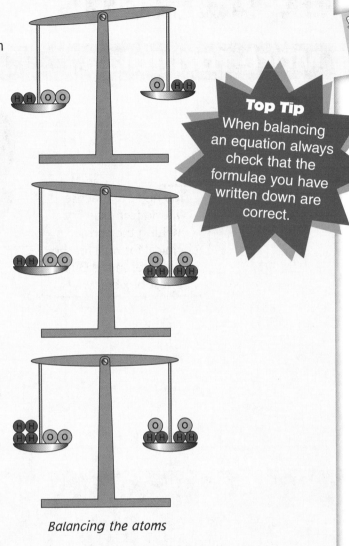

Balancing the atoms

Top Tip
When balancing an equation always check that the formulae you have written down are correct.

Top Tip
If you have to write the equation for a reaction, it may be easier to write it in words first.

CREDIT

Quick Test

1. How many calcium atoms are present in $CaCO_3$?

2. How many carbon atoms are present in $CaCO_3$?

3. How many oxygen atoms are present in $CaCO_3$?

4. Why must there be the same number of atoms on both sides of the equation?

5. Balance the equation
   $Na(s) + Cl_2(g) \rightarrow NaCl(s)$.

6. Balance the equation
   $H_2(g) + Cl_2(g) \rightarrow HCl(g)$.

7. Balance the equation
   $C(s) + CO_2(g) \rightarrow CO(g)$.

8. What does the state symbol (l) indicate?

9. What does the state symbol (aq) indicate?

10. Add the state symbols to this equation for the thermal decomposition of calcium carbonate:
    $CaCO_3 \rightarrow CaO + CO_2$

Answers 1. 1 2. 1 3. 3 4. Atoms cannot be created or destroyed. 5. $2Na(s) + Cl_2(g) \rightarrow 2NaCl(s)$ 6. $H_2(g) + Cl_2(g) \rightarrow 2HCl(g)$ 7. $C(s) + CO_2(g) \rightarrow 2CO(g)$ 8. Liquid 9. Aqueous 10. $CaCO_3(s) \rightarrow CaO(s) + CO_2(g)$

More formulae and equations

More formulae

You need to practise writing formulae. Try writing the correct chemical formula for the substances in the Quick Test box below:

Quick Test

1. oxygen gas
2. silver(I) nitrate *CREDIT*
3. magnesium fluoride
4. calcium sulphide
5. calcium sulphate
6. sodium sulphide
7. sodium sulphite
8. calcium hydroxide *CREDIT*
9. magnesium nitrate
10. aluminium sulphate *CREDIT*
11. nickel(II) oxide *CREDIT*
12. nickel(II) chloride *CREDIT*
13. carbon monoxide
14. sulphur trioxide
15. nitrogen dioxide
16. phosphorus trichloride
17. lead(II) carbonate *CREDIT*
18. silicon dioxide
19. lead(II) bromide *CREDIT*
20. aluminium hydroxide *CREDIT*
21. sodium sulphate
22. magnesium phosphate *CREDIT*
23. ammonium sulphite *CREDIT*
24. potassium iodide
25. lithium nitrate
26. lithium sulphate
27. nitrogen trichloride

Answers 1. O_2, 2. $AgNO_3$ 3. MgF_2 4. CaS 5. $CaSO_4$ 6. Na_2S 7. Na_2SO_3 8. $Ca(OH)_2$ 9. $Mg(NO_3)_2$ 10. $Al_2(SO_4)_3$ 11. NiO 12. $NiCl_2$ 13. CO 14. SO_3 15. NO_2 16. PCl_3 17. $PbCO_3$ 18. SiO_2 19. $PbBr_2$ 20. $Al(OH)_3$ 21. Na_2SO_4 22. $Mg_3(PO_4)_2$ 23. $(NH_4)_2SO_3$ 24. KI 25. $LiNO_3$ 26. Li_2SO_4 27. NCl_3

More equations

Write formula equations, and balance them where required, for the word equations in the Quick Test box below :

Quick Test

1. magnesium + oxygen → magnesium oxide

2. calcium + sulphur → calcium sulphide

3. sodium hydroxide + hydrochloric acid → sodium chloride + water

4. sodium + oxygen → sodium oxide

5. sodium + water → sodium hydroxide + hydrogen

6. calcium + water → calcium hydroxide + hydrogen

7. magnesium + hydrochloric acid → magnesium chloride + hydrogen

8. aluminium + hydrochloric acid → aluminium chloride + hydrogen

9. calcium carbonate + sulphuric acid → calcium sulphate + water + carbon dioxide

10. calcium + oxygen → calcium oxide

11. potassium hydroxide + nitric acid → potassium nitrate + water

12. potassium + oxygen → potassium oxide

13. aluminium + oxygen → aluminium oxide

14. nitric acid + magnesium carbonate → magnesium nitrate + water + carbon dioxide

15. propane + oxygen → carbon dioxide + water

16. calcium + nitric acid → calcium nitrate + hydrogen

17. ethene + oxygen → carbon dioxide + water

18. carbon monoxide + oxygen → carbon dioxide

Answers 1. $2Mg + O_2 \rightarrow 2MgO$ **2.** $Ca + S \rightarrow CaS$ **3.** $NaOH + HCl \rightarrow NaCl + H_2O$ **4.** $4Na + O_2 \rightarrow 2Na_2O$ **5.** $2Na + 2H_2O \rightarrow 2NaOH + H_2$ **6.** $Ca + 2H_2O \rightarrow Ca(OH)_2 + H_2$ **7.** $Mg + 2HCl \rightarrow MgCl_2 + H_2$ **8.** $2Al + 6HCl \rightarrow 2AlCl_3 + 3H_2$ **9.** $CaCO_3 + H_2SO_4 \rightarrow CaSO_4 + H_2O + CO_2$ **10.** $2Ca + O_2 \rightarrow 2CaO$ **11.** $KOH + HNO_3 \rightarrow KNO_3 + H_2O$ **12.** $4K + O_2 \rightarrow 2K_2O$ **13.** $4Al + 3O_2 \rightarrow 2Al_2O_3$ **14.** $2HNO_3 + MgCO_3 \rightarrow Mg(NO_3)_2 + H_2O + CO_2$ **15.** $C_3H_8 + 5O_2 \rightarrow 3CO_2 + 4H_2O$ **16.** $Ca + 2HNO_3 \rightarrow Ca(NO_3)_2 + H_2$ **17.** $C_2H_4 + 3O_2 \rightarrow 2CO_2 + 2H_2O$ **18.** $2CO + O_2 \rightarrow 2CO_2$

Ionic bonding

Ionic bonding involves the transfer of electrons.

How ions are formed

All atoms become more stable if they have a **full outer shell** of electrons (like the noble gases).

Ionic bonding involves the **transfer** of electrons from one atom to another. Metals in Groups 1 and 2 – such as sodium and calcium – lose negative electrons to gain a full outer shell. Overall they become positively charged (electrons are negative).

Non-metals in Groups 6 and 7 – such as oxygen and chlorine – gain negative electrons to attain a full outer shell. So overall they become negatively charged.

The table shows the ions formed by the first 20 elements.

H^+ hydrogen									Group 0 None helium
Group 1	**Group 2**		**Group 3**	**Group 4**	**Group 5**	**Group 6**	**Group 7**		
Li^+ lithium	Be^{2+} beryllium		**None** boron	**None** carbon	N^{3-} nitride	O^{2-} oxide	F^- fluoride		**None** neon
Na^+ sodium	Mg^{2+} magnesium		Al^{3+} aluminium	**None** silicon	P^{3-} phosphide	S^{2-} sulphide	Cl^- chloride		**None** argon
K^+ potassium	Ca^{2+} calcium	transition metals							

In neutral atoms there are the **same number** of **positive protons** as there are **negative electrons**. This means that if electrons are lost or gained the number of **protons** and the number of **electrons** is no longer **balanced**. So an ion is an **atom**, or a **small group** of atoms, with a **charge**.

Top Tip
When drawing the dot and cross diagrams to show ionic or covalent bonding, just draw the outer shells.

Examples of ionic compounds

sodium + chlorine → sodium chloride

Na + $\frac{1}{2}Cl_2$ → Na^+Cl^-

Note: the dots (●) and crosses (×) represent electrons.

The sodium atom **transfers** an electron to the chlorine atom. Both the sodium and the chlorine atoms now have full outer shells. Sodium has lost a negative electron so becomes **positively** charged. Chlorine has **gained** an electron so becomes **negatively** charged. The positive sodium ions are attracted to the negative chloride ions and the ionic compound sodium chloride is made.

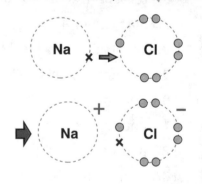

sodium + fluorine → sodium fluoride

Na + $\frac{1}{2}F_2$ → Na^+F^-

Fluorine and chlorine are in the same group of the periodic table, Group 7. Fluorine and sodium will therefore react in a similar way to chlorine and sodium. Sodium transfers one electron to fluorine. The positive sodium ions and negative fluoride ions attract each other and the ionic compound sodium fluoride is formed.

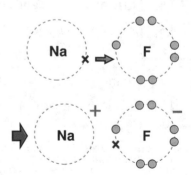

More examples of ionic compounds

Note: the dots (●) and crosses (×) represent electrons.

magnesium + oxygen → magnesium oxide

$$Mg \quad + \quad \tfrac{1}{2}O_2 \quad \rightarrow \quad Mg^{2+}O^{2-}$$

Magnesium transfers two electrons to oxygen. The magnesium atoms become magnesium ions with a 2+ charge. The oxygen atoms become oxide ions with a 2− charge.

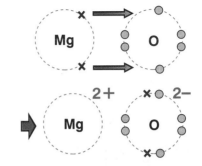

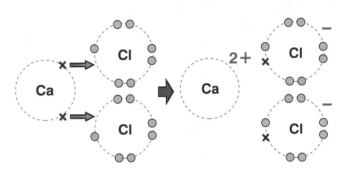

calcium + chlorine → calcium chloride

$$Ca \quad + \quad Cl_2 \quad \rightarrow \quad Ca^{2+}(Cl^-)_2$$

Calcium transfers two electrons in total, one to each chlorine atom. Each calcium ion formed has a 2+ charge. Each chloride ion formed has a 1− charge, so two chloride ions are needed for every calcium ion so that the charges are balanced.

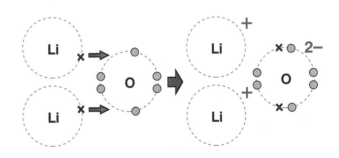

lithium + oxygen → lithium oxide

$$Li \quad + \quad O_2 \quad \rightarrow \quad (Li^+)_2O^{2-}$$

Each lithium atom transfers one electron to the oxygen atom, forming Li^+ and O^{2-} ions.

All these compounds are held together by strong forces of attraction between the oppositely charged ions. Ionic compounds usually contain metal elements joined to non-metal elements.

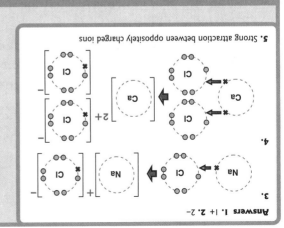

Quick Test

1. If an element in Group 1 loses an electron, what charge does it have?
2. If an element in Group 6 gains two electrons, what charge does it have?
3. Draw a dot and cross diagram to show sodium reacting with chlorine.
4. Draw a dot and cross diagram to show calcium reacting with chlorine.
5. What holds together ionic compounds?

Answers 1. 1+ 2. 2−
3.
4.
5. Strong attraction between oppositely charged ions

Structures of ionic and covalent compounds

Ionic bonding occurs between **metals** in Groups 1 and 2 and **non-metals** in Groups 6 and 7. It involves the **transfer of electrons**. **Covalent bonding** occurs between non-metal atoms. It involves the **sharing of electrons**.

Structures of ionic compounds

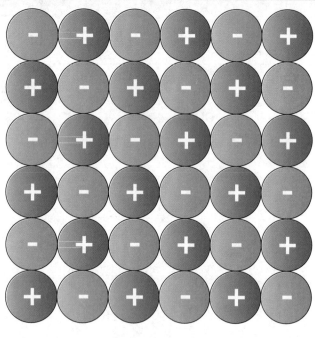

An ionic lattice

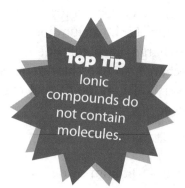

Top Tip
Ionic compounds do not contain molecules.

- The ions are held together in a **regular structure** known as a **lattice.**

- The **strong forces of attraction** between **oppositely charged ions** means that they have **very high melting and boiling points** and so are **solids** at room temperature.

- When ionic compounds dissolve in water, the **lattice is broken** and the ions are now **free to move**.

- The **lattice is also broken** when the ionic compound is heated until it **melts.**

- In the **solid state the ions are trapped in the lattice**. The **ions** can **move** around **in solution** and also when the ionic compound has been **melted**.

Top Tip
Pure ionic compounds are always solid at room temperature

Structures of covalent compounds

Atoms which **share electrons** form **molecules.** Inside a molecule the atoms are held together by **shared pairs of electrons**. A molecule contains two or more atoms held together by covalent bonds.

Examples of simple molecular covalent structures

Oxygen, **water**, **ammonia**, **methane** and **carbon dioxide**

oxygen

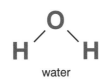
water

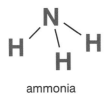
ammonia

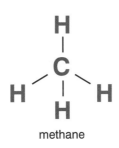
methane

$$O = C = O$$
carbon dioxide

These molecules are formed from small numbers of atoms.

There are strong covalent bonds between the atoms in each molecule, but very weak forces of attraction between these molecules.

This means that simple molecular compounds have low melting and boiling points.

Most are gases or liquids at room temperature.

Substances which are liquid or gas at room temperature have covalent bonding (except for mercury)

Simple molecular substances do not conduct electricity because they do not contain ions.

They tend to be insoluble in water (although they may dissolve in other solvents such as petrol).

Covalent substances which are **solid** at room temperature have **larger molecules**. **Covalent network** substances such as **diamond or silicon dioxide** have **very high melting points**.

Top Tip
Covalent compounds can be solids, liquids or gases at room temperature depending on the size of their molecules.

Quick Test

1. What kind of structure do ionic compounds form?

2. Why do ionic compounds have high melting and boiling points?

3. In which states are ions free to move?

4. Give an example of a simple molecular substance.

5. In simple molecular compounds describe the bonding between atoms and between molecules.

6. Do simple molecular compounds conduct electricity?

Answers 1. Giant regular lattice structures **2.** Because of the strong forces of attraction between oppositely charged ions. **3.** Ions can move when dissolved and when they are molten. **4.** Oxygen/water/ammonia/methane/carbon dioxide; etc. **5.** Strong attraction between atoms, weak between molecules **6.** No

More about ionic compounds

Conducting electricity

When ionic compounds are solid the ions are trapped in a regular structure known as a **lattice.** In this state the **ions are not** free to move and so the ionic compound **cannot conduct electricity**.

For the ionic compound to conduct electricity, the **lattice must be broken** so that the **ions are free to move.** This can be done in two ways:

- the ionic compound can be melted into a liquid
- the ionic compound can be dissolved in water to make an aqueous solution.

Electrolysis

A solution containing ions is an **electrolyte. Electrolysis** is when electricity is passed through an ionic solution. When an ionic compound is electrolysed, the **positive metal ions** move to the **negative electrode** where they pick up electrons and change into metal atoms. This **gain of electrons** is known as **reduction**.

At the same time, the **negative non-metal ions** travel to the **positive electrode** where they lose electrons and change into molecules. This **loss of electrons** is known as **oxidation.**

When ionic compounds conduct electricity the ionic compound can change or undergo **decomposition.**

For example, when copper chloride solution is electrolysed it changes into copper and chlorine. We say that the copper chloride has **decomposed** into copper and chlorine.

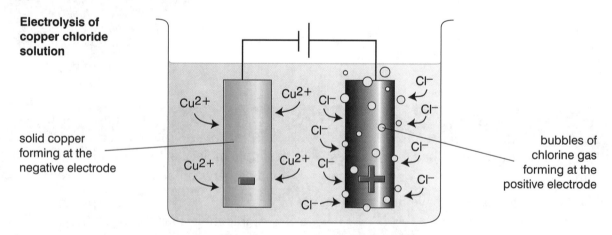

Electrolysis of copper chloride solution

solid copper forming at the negative electrode

Cu^{2+} Cu^{2+} Cl^- Cl^-

Cl^- Cl^-

Cu^{2+} Cu^{2+} Cl^- Cl^-

$Cl^- \rightarrow$

bubbles of chlorine gas forming at the positive electrode

Electrolysis of copper chloride solution

Since positive and negative electrodes are needed in this type of experiment, a d.c. supply of electricity must be used if the products are to be identified.

CREDIT

Ion-electron equations

CREDIT

Equations known as **ion-electron equations** are used to describe what happens during electrolysis.

- What happens at the positive electrode during the electrolysis of copper chloride is shown in the equation:

$$2Cl^-(aq) \rightarrow Cl_2(g) + 2e^-$$

The negative chloride ions have travelled to the positive electrode and lost electrons and have changed into chlorine gas. **A loss of electrons is oxidation**.

- What happens at the negative electrode during the electrolysis of copper chloride is shown in the equation:

$$Cu^{2+}(aq) + 2e^- \rightarrow Cu(s)$$

The positive copper ions have travelled to the negative electrode and gained electrons to change into copper atoms. **A gain of electrons is reduction.**

Ion-electron equations are difficult to write, but all the equations you need are to be found in the Data Booklet page 7 where they are all written in reduction form. Oxidation equations are written by reversing the reduction equation.

Top Tip
Try to remember **OILRIG** which means **O**xidation **I**s **L**oss and **R**eduction **I**s **G**ain.

Quick Test

1. In which two states do ionic compounds conduct electricity?

2. Why do ionic compounds not conduct electricity in the solid state?

3. What is meant by reduction?

4. What is meant by oxidation?

5. Why do copper ions travel to the negative electrode during the electrolysis of copper chloride solution?

6. When molten lead bromide, $PbBr_2$, is electrolysed it decomposes into lead and bromine. Using your Data Booklet to help you, write the two ion-electron equations for the reaction.

Answers 1. In solution and when molten. 2. The ions are not free to move. 3. A gain of electrons. 4. A loss of electrons. 5. Because the copper ions are positively charged. 6. $Pb^{2+} + 2e^- \rightarrow Pb$; $2Br^- \rightarrow Br_2 + 2e^-$

Ionic compounds: colours and formulae

Coloured ions

Most ionic compounds are white when solid and dissolve in water to give colourless solutions. However, ions containing **transition metals** are often coloured. The table below shows some examples.

Row	Name of compound	Positive ions present	Negative ions present	Colour
1	sodium chloride	sodium	chloride	colourless
2	copper chloride	copper	chloride	blue
3	sodium chromate	sodium	chromate	yellow
4	potassium nitrate	potassium	nitrate	colourless
5	potassium permanganate	potassium	permanganate	purple
6	sodium sulphate	sodium	sulphate	colourless
7	nickel sulphate	nickel	sulphate	green
8	copper chromate	copper	chromate	green

- If we look at row 1 in the table we can see both sodium and chloride ions must be colourless.
- Remembering that chloride ions are colourless and looking at row 2, we can see that copper ions must be blue.
- If we now look at row 3, it should be obvious that chromate ions must be yellow.
- Copper chromate is coloured green because it contains blue copper ions and yellow chromate ions.

Ion migration

Ion migration experiments such as the one shown below demonstrate that the blue copper ions travel to the negative electrode and the yellow chromate ions travel to the positive electrode. This confirms that copper ions are positively charged and that chromate ions are negatively charged.

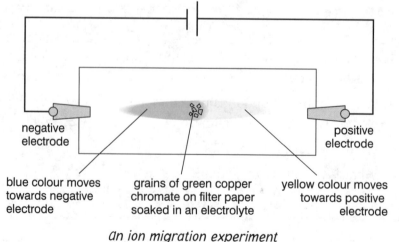

negative electrode

positive electrode

blue colour moves towards negative electrode

grains of green copper chromate on filter paper soaked in an electrolyte

yellow colour moves towards positive electrode

an ion migration experiment

Ionic compounds: colours and formulae

Formulae of ions

Metal ions have **positive** charges. The amount of charge depends on the valency of the metal and this depends on the group that the metal belongs to in the Periodic Table. For example, sodium is in Group 1, magnesium is in Group 2 and aluminium is in Group 3 so their ions are Na^+, Mg^{2+} and Al^{3+} respectively.

Ions of non-metal elements usually have negative charges and the amount of charge also depends on the valency of the element. For example, oxygen is a non-metal and has valency = 2, so the oxide ion has a charge of 2– and the formula is O^{2-}. Similarly the chloride ion is Cl^- and the nitride ion is N^{3-}.

Top Tip
Use your Data Booklet to help you write ionic formulae correctly. Pages 4 and 7 give formulae of different ions.

Formulae of ionic compounds

When we write the ionic formulae for compounds the total number of positive charges must be balanced by the total number of negative charges.

- In sodium chloride the sodium ion is Na^+ and this positive charge is balanced by the Cl^- ion. The ionic formula for sodium chloride is Na^+Cl^-.
- The ionic formula for magnesium oxide is $Mg^{2+}O^{2-}$. The 2+ charge on the magnesium ion balances the 2– charge on the oxide ion.
- For sodium oxide the sodium ion is Na^+ and the oxide ion is O^{2-}. Two sodium ions are needed to balance the one oxide ion and the formula is written as $(Na^+)_2O^{2-}$.

Top Tip
You can only write ionic formulae for ionic compounds, not for covalent substances. For example, there is no ionic formula for carbon dioxide.

Quick Test

1. Look at the table on page 36 and deduce the colours of the following ions:
   a) potassium
   b) permanganate
   c) nickel.

2. Predict the colour of nickel chloride.

3. Why does the yellow colour move towards the positive electrode in the ion migration experiment?

4. Write the formulae of the following ions:
   a) potassium
   b) sulphide
   c) fluoride
   d) aluminium.

5. Write ionic formulae for the following compounds:
   a) sodium fluoride
   b) magnesium sulphide
   c) aluminium oxide
   d) calcium chloride
   e) calcium hydroxide.

Answers 1. a) colourless **b)** purple **c)** green. **2.** Green. **3.** Because the yellow chromate ions are negatively charged. **4. a)** K^+ **b)** S^{2-} **c)** F^- **d)** Al^{3+} **5. a)** Na^+F^- **b)** $Mg^{2+}S^{2-}$ **c)** $(Al^{3+})_2(O^{2-})_3$ **d)** $Ca^{2+}(Cl^-)_2$ **e)** $Ca^{2+}(OH^-)_2$

37

Fuels

What is a fuel?

A **fuel** is a **chemical** which **burns** to give out **energy.**

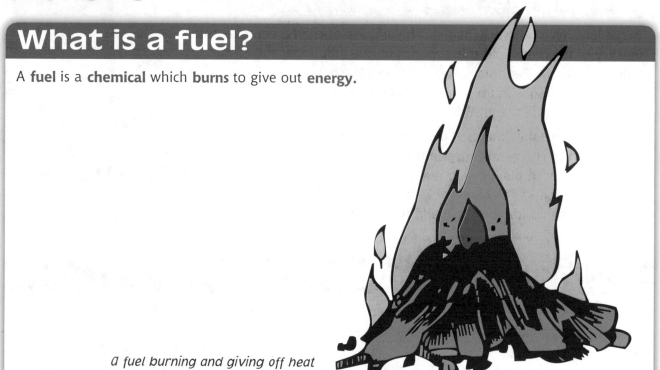

a fuel burning and giving off heat

Combustion

When a fuel burns the chemical reaction is known as **combustion**. When combustion takes place the fuel is **reacting with oxygen** from the air and **energy is given out**. This means that combustion is an example of an **exothermic reaction**.

Composition of air

Air is a **mixture** of different gases. Nitrogen is the main component of air. Apart from oxygen, the other gases are mainly argon and carbon dioxide. Air usually contains water vapour as well as these other gases.

Gas	% of dry air
Nitrogen	78
Oxygen	21
Other gases	1

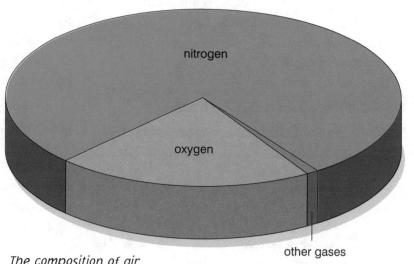

The composition of air

Tests for gases

Oxygen

Oxygen **relights** a glowing **splint**.

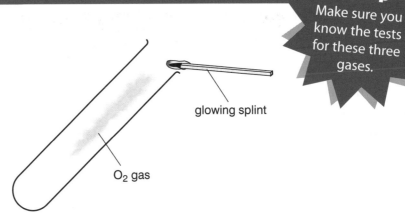

glowing splint

O₂ gas

Carbon dioxide

The gas is **bubbled** through **limewater**.
Carbon dioxide turns limewater **cloudy**.

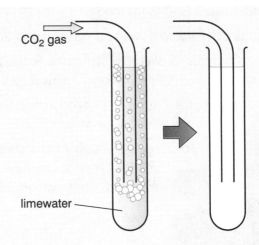

CO₂ gas

limewater

Hydrogen

If a lighted splint is nearby hydrogen will burn with a '**squeaky pop**'.

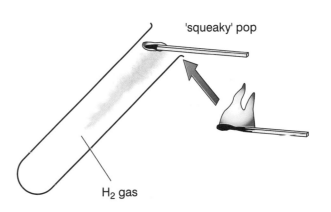

'squeaky' pop

H₂ gas

Quick Test

1. What is a *fuel*?

2. What does *combustion* mean?

3. What is meant by an *exothermic* reaction?

4. What percentage of dry air is oxygen?

5. Which gas turns lime water cloudy?

6. Which gas burns with a squeaky pop?

7. What is the test for oxygen gas?

Answers 1. A chemical which burns to give out energy. 2. Burning. 3. A reaction which gives out energy. 4. 21% 5. Carbon dioxide. 6. Hydrogen. 7. It re-lights a glowing splint.

Fossil fuels

Formation of coal, oil and natural gas

Coal, oil and natural gas are fossil fuels.

Fossil fuels are formed over **millions of years**. They are the fossilised remains of dead plants and animals.

Plants and animals died and fell to the sea or swamp floor.

The remains were quickly covered by **sediment**.

In the absence of **oxygen** the remains did not **decay**.

As the layers of sediment **increased** the remains became **heated and pressurised** (squashed).

After millions of years **coal**, **oil** and **natural gas** are formed.

Dead plants falling into swamps form **coal**, while tiny dead sea creatures and plants form **oil and natural gas**.

Burning fossil fuels is an exothermic process – giving out a lot of heat.

Fossil fuels are **non-renewable**. They take millions of years to form, but they are being used up very quickly.

Top Tip
Know how fossil fuels are formed and that the original source of their energy is the Sun.

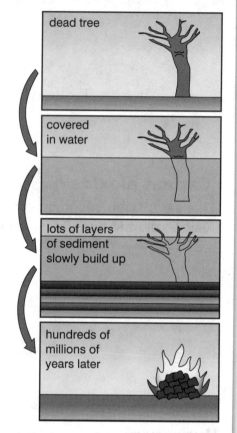

dead tree

covered in water

lots of layers of sediment slowly build up

hundreds of millions of years later

The build up of dead plant material to form fossil fuels

Fuel crisis

As the human population of the world continues to grow rapidly, fossil fuels are being used up at an alarming rate. Fossil fuels are a **finite resource**. Once they have been used up they **cannot be replaced** as fossil fuels take **hundreds of millions of years to form**. It is difficult to estimate how long fossil fuels will last. **Coal** may last for **300 years**, but **oil and natural gas** are expected to run out in **40–70 years.** To reduce this **fuel crisis**, research in finding and using **alternative fuels** is being carried out. For example, in Brazil **alcohol from sugar cane** is being used as an **alternative to petrol**.

Pollution of the atmosphere

- The burning of fossil fuels can affect the atmosphere.
- Most fuels contain both **carbon** and **hydrogen** and many also contain a little **sulphur**.
- When **carbon** is **burned**, **carbon dioxide** (CO_2) is released into the atmosphere.
- When **hydrogen** is **burned**, **water vapour** (H_2O) is released (water is an **oxide** of hydrogen).
- When **sulphur is burned**, **sulphur dioxide** (SO_2) is formed. Removing sulphur and sulphur compounds from fossil fuels before they are burned reduces air pollution.

Acid rain

- Fossil fuels may contain some **sulphur**.
- When these fuels are burned **sulphur dioxide** is produced and released into the atmosphere.
- This gas dissolves in rainwater to produce **acid rain**.
- Acid rain can affect the environment, damaging statues and buildings as well as trees, animals and plants.

Acid rain kills trees

Top Tip
Remember that SO_2 is the main cause of acid rain and make sure you know about the damage it causes.

Carbon dioxide and the greenhouse effect

Top Tip
CO_2 is the main cause of the greenhouse effect.

- The **greenhouse effect** is slowly heating up the Earth.
- When fossil fuels are burned, **carbon dioxide** is produced.
- Although some of this carbon dioxide is removed from the atmosphere when the gas dissolves in the oceans, the overall amount of carbon dioxide in the atmosphere has **gradually increased** over the last 200 years.
- This carbon dioxide **traps the heat** that has reached the Earth from the Sun.
- Global warming may mean that the ice at the North and South Poles will melt and cause massive flooding.

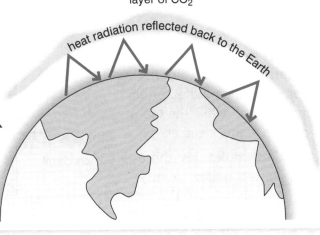

layer of CO_2

heat radiation reflected back to the Earth

light energy from the Sun

The greenhouse effect

Quick Test

1. What is meant by 'a finite resource'?

2. Name the compounds formed when hydrogen, carbon and sulphur are burned.

3. Which gas is thought to be responsible for
   a) the greenhouse effect   **b)** acid rain?

4. Give two examples of the damaging effects of acid rain.

Answers 1. One that is running out and cannot be replaced. **2.** Water, carbon dioxide and sulphur dioxide. **3. a)** CO_2 **b)** SO_2 **4.** It erodes buildings and statues, corrodes metals and kills plant and animal life in rivers and lochs.

Crude oil

What is crude oil?

Crude oil is a mixture of different **hydrocarbons**. A **hydrocarbon** is a compound containing **only hydrogen and carbon**. The different hydrocarbons present in crude oil have different melting and boiling points. Because these hydrocarbon compounds have different boiling points, they can be separated into **fractions** using **fractional distillation**. A **fraction** is a group of compounds with boiling points within a definite range.

Top Tip
Remember that hydrocarbons contain **only** hydrogen and carbon.

Fractional distillation of crude oil

- Crude oil is a **mixture** of **hydrocarbons**.
- Hydrocarbons are molecules that contain only **hydrogen** and **carbon atoms**.
- Some of the hydrocarbons have **very short chains** of carbon atoms.
- These hydrocarbons are **runny**, **easy to ignite** and have **low boiling points**.
- Other hydrocarbons have much longer chains of carbon atoms. These hydrocarbons are more **viscous** (less runny), **harder to ignite** and have **higher boiling points**.
- This means that long chain hydrocarbons are **not useful as fuels**.

Fractional distillation can be used to separate mixtures of **hydrocarbons (crude oil)**. In the fractionating column the bottom is kept very hot, while the top of the column is much cooler.

The **smallest molecules** (those with the lowest boiling points) boil off first and **rise to the top of the column**. The other **fractions** are collected at different points down the column.

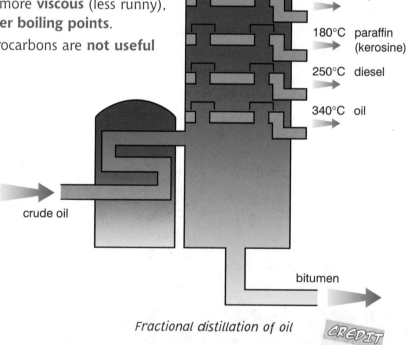

Fractional distillation of oil

fuel gas

40°C petrol (gasoline)

110°C naphtha

180°C paraffin (kerosine)

250°C diesel

340°C oil

crude oil

bitumen

CREDIT

The **fraction** which collects **at the top of the column** has the **lowest boiling point** and so evaporates easiest. It is also the most flammable fraction and is the least viscous. **Bitumen** is the most difficult to evaporate and collects at the bottom of the column. It is the least flammable and the most viscous (thickest liquid). The **fraction** which collects **at the top** of the column has the **smallest molecules** whereas the biggest molecules are present in bitumen. **Bigger molecules have higher boiling points** than smaller molecules as they require more energy to move about rapidly.

Fractions and their uses

The actual composition of the crude oil will depend on the source of the oil. For example, North Sea oil will be slightly different from South American oil. The table below shows a possible result from fractional distillation.

Fraction	Boiling point range (°C)	Number of carbon atoms	Uses
Fuel gases	−160 to 20	1–4	Fool for cooking and heating
Petrol (gasoline)	20 to 75	5–10	Fuel for cars
Naphtha	75 to 130	8–12	Chemicals
Paraffin (kerosene)	130 to 240	10–16	Aviation fuel, heating
Diesel oil	240 to 300	15–28	Fuel for diesel engines in lorries, trains and other vehicles
Lubricating oils and waxes	300 to 350	26–70	Lubricating oils, polishes
Bitumen	Above 350	More than 70	Roads and roofing

Top Tip
You should know a use for each of the fractions.

Quick Test

1. What is meant by a *hydrocarbon*?

2. What is a *fraction*?

3. Why can the different hydrocarbons in crude oil be separated by fractional distillation?

4. Which fraction is the most flammable?

5. Which fraction is the most viscous?

6. Which are smaller, molecules in petrol or molecules in paraffin?

7. What is diesel oil used for?

8. What is bitumen used for?

9. The hydrocarbon, butane, has the molecular formula C_4H_{10}. In which fraction will butane be found?

10. Which fraction has the smallest molecules?

Answers 1. A compound containing **only** hydrogen and carbon. **2.** A group of compounds with similar boiling points. **3.** Because they have different boiling points. **4.** Fuel gases. **5.** Bitumen. **6.** Molecules in petrol. **7.** Fuel for trains. **8.** Roads and roofing. **9.** Fuel gases. **10.** Fuel gases.

Combustion of hydrocarbons

Complete combustion

When **hydrocarbons burn** they are reacting with oxygen in the air. If there is an **adequate supply of oxygen** the **hydrocarbons burn completely** (complete combustion) producing **carbon dioxide and water.** This can be demonstrated by carrying out the experiment shown below.

Top Tip
Make sure you can draw this diagram accurately. Look carefully at the lengths of the tubing inside the test tubes.

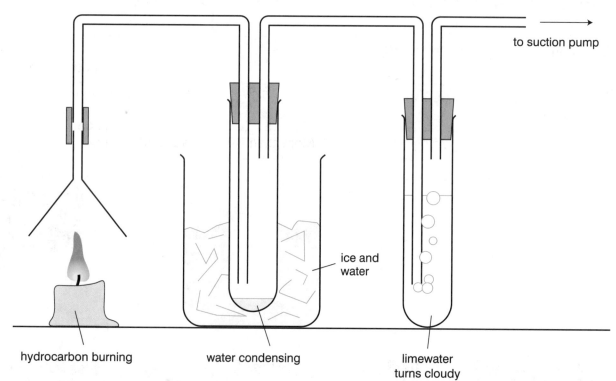

to suction pump

ice and water

hydrocarbon burning   water condensing   limewater turns cloudy

Experiment to demonstrate the effects of burning hydrocarbons

- The liquid collected in the first test tube **freezes at 0 ºC** and **boils at 100 ºC** and so must be **water.**
- The **lime water** in the second test tube turns **cloudy** and so **carbon dioxide** must also have been formed by the burning hydrocarbon.
- As carbon dioxide is produced we can say that there must be carbon present in the fuel. The carbon in the carbon dioxide must have come from the fuel.
- As water is produced we can say that there must be hydrogen present in the fuel. The hydrogen in the water must have come from the fuel.
- The oxygen present in the carbon dioxide and in the water will have come from the air.

Incomplete combustion

If there is not enough oxygen present, then, as well as water forming, the other products of combustion may be **carbon (soot)** and the very poisonous **carbon monoxide (CO)** instead of carbon dioxide. This is known as **incomplete combustion.**

The carbon monoxide formed leads to another form of **air pollution**. Carbon monoxide stops the blood carrying oxygen around the body and so is very toxic.

Reducing air pollution

There is a limited supply of oxygen inside a car engine and the burning petrol may produce carbon monoxide. To reduce the amount of carbon monoxide going into the air, car manufacturers now install a **catalytic converter** into the exhaust systems of cars. This converts the poisonous carbon monoxide into the less harmful carbon dioxide.

CREDIT

> Increasing the air to fuel ratio (or decreasing the fuel to air ratio) allows more complete combustion to take place inside petrol engines. The more air there is present compared to petrol then the less chance of carbon monoxide being produced and the more efficient the combustion.

A hot spark from the spark plug ignites the petrol inside the engine. The energy from this spark enables the nitrogen in the air to react with the oxygen. The result of this reaction is a mixture of different oxides of nitrogen such as NO and NO_2. These are also poisonous gases. The catalytic converter will turn these gases back into harmless nitrogen and oxygen. The catalytic converter will also change any hydrocarbons that have not burned inside the engine into harmless water and carbon dioxide.

Petrol used to contain compounds of lead so that it would burn smoothly inside car engines. These lead compounds also caused air pollution, so petrol is now lead free.

Top Tip
Remember that a catalytic converter turns pollutant gases into harmless gases.

Quick Test

1. What are the products when a hydrocarbon burns in a plentiful supply of air?

2. What are the products of incomplete combustion of a hydrocarbon fuel?

3. What is the test for water?

4. Write the chemical formula for carbon monoxide.

5. Draw a diagram to show that a burning candle produces carbon dioxide and water.

6. What causes nitrogen and oxygen in air to combine inside a car engine?

7. What does a catalytic converter do?

8. How does changing the fuel to air ratio lead to less carbon monoxide being produced?

Answers 1. CO_2 and H_2O **2.** Carbon (soot); carbon monoxide and water. **3.** It freezes at 0 °C and boils at 100 °C. **4.** CO **5.** See diagram opposite. **6.** The energy from the spark which is used to ignite the petrol. **7.** It converts pollutant gases into harmless gases. **8.** Decreasing the fuel to air ratio means that there will be more oxygen compared to fuel and so more complete combustion will take place.

Alkanes

The alkane family

The **alkanes** are a series of **hydrocarbons**. This means that the alkanes contain hydrogen and carbon only.

All hydrocarbons have a **spine** of **carbon atoms**.

Molecules that belong to the **alkane** family contain **no carbon-to-carbon double bonds**. They are **saturated hydrocarbons**. All the **carbon-to-carbon bonds** in a **saturated** hydrocarbon are **single** bonds. The names of all the members of the alkane family end in **-ane**

The smaller alkanes are gases, but as the molecules increase in size they become liquids and eventually solids. Page 6 of the Data Booklet gives the melting and boiling points of the first eight alkanes.

Top Tip
The names of the first eight alkanes are given in the Data Booklet page 6.

Homologous series

The alkane family make up a **homologous series**. This means that:

- they are all covered by the **same general formula** (C_nH_{2n+2})
- there is a **constant increment** in their formula between one member of the series and the next (CH_2)
- they all have **similar chemical properties** (take part in the same chemical reactions)
- there is a **gradation in their physical properties**, such as melting and boiling points.

As the number of carbon atoms increases in the formula then the molecules become bigger in size. The bigger the molecules, the more energy is required to get the molecules to move about rapidly. This explains why the alkanes with the smallest molecules have the lowest boiling points and those with the largest molecules have higher boiling points

Molecular and structural formulae

- The **molecular formula** tells you the **number** of carbon and hydrogen **atoms in each molecule**.
- The **structural formula** shows the **bonds** holding the atoms together. It is simplified so that the molecules look flat, but the atoms are arranged **tetrahedrally.**

The simplest alkane is **methane**. It has molecular formula CH_4. This means that every methane molecule contains **one carbon atom and four hydrogen atoms**. The structural formula of methane shows that the carbon atom is **covalently** bonded to each of the four hydrogen atoms.

$$\begin{array}{c} H \\ | \\ H-C-H \\ | \\ H \end{array}$$

Formulae of the alkanes

The names, molecular formula and structural formula of the alkanes are shown in the table below. Each carbon atom forms four covalent bonds. Each hydrogen atom can only form one covalent bond.

Top Tip
Make sure you know the names, molecular formulae and can draw the structures of these alkanes.

Name	Molecular formula	Structural formula	Shortened structural formula
Methane	CH_4	H–C–H (with H above and below)	CH_4
Ethane	C_2H_6	H–C–C–H	CH_3CH_3
Propane	C_3H_8	H–C–C–C–H	$CH_3CH_2CH_3$
Butane	C_4H_{10}	H–C–C–C–C–H	$CH_3CH_2CH_2CH_3$
Pentane	C_5H_{12}	H–C–C–C–C–C–H	$CH_3CH_2CH_2CH_2CH_3$
Hexane	C_6H_{14}	H–C–C–C–C–C–C–H	$CH_3CH_2CH_2CH_2CH_2CH_3$
Heptane	C_7H_{16}	H–C–C–C–C–C–C–C–H	$CH_3CH_2CH_2CH_2CH_2CH_2CH_3$
Octane	C_8H_{18}	H–C–C–C–C–C–C–C–C–H	$CH_3CH_2CH_2CH_2CH_2CH_2CH_2CH_3$

Quick Test

1. What is meant by a *saturated hydrocarbon*?
2. Write the general formula for the alkane series.
3. What does the molecular formula tell us?
4. Write the molecular formula and draw the structural formula for:
   a) methane
   b) butane.

Answers 1. A hydrocarbon that has no C=C double bonds or one in which all the C–C bonds are single. 2. C_nH_{2n+2} 3. The number of atoms of each element in a molecule. 4. a) CH_4 b) C_4H_{10}

Alkenes

The alkene family

The **alkenes** are another series of **hydrocarbons**. Alkenes also contain hydrogen and carbon only. Like the alkanes, the alkenes have a spine of carbon atoms.

The alkene hydrocarbons contain **carbon-to-carbon double bonds**. They are **unsaturated hydrocarbons** as they contain a double carbon-to-carbon bond. The **names** of all the members of the alkene family end in **-ene**

Alkenes generally have lower melting and boiling points than their equivalent alkanes. The smaller alkenes are gases, but as the molecules increase in size they become liquids and eventually solids. Page 6 in the Data Booklet gives the melting and boiling points of the first five alkenes.

As alkenes must have a double carbon to carbon bond, the simplest alkene must have **two carbon atoms**. The **first member** of the alkene family is called **ethene.** The alkene family also make up a **homologous series**. The general formula for the alkene family is C_nH_{2n}.

> **Top Tip**
> The names of the first five alkenes are given in the Data Booklet page 6.

Formulae of the alkenes

> **Top Tip**
> Make sure you know the names, molecular formulae and can draw the structures of these alkenes.

The names, molecular formulae and structural formulae of the first five members of the alkene family are shown in the table below.

Name	Molecular formula	Structural formula	Shortened structural formula
Ethene	C_2H_4	$\begin{array}{cc} H & H \\ | & | \\ C&=&C \\ | & | \\ H & H \end{array}$	$CH_2=CH_2$
Propene	C_3H_6	$\begin{array}{ccc} H & H & H \\ | & | & | \\ C&=&C&-&C&-&H \\ | & & | \\ H & & H \end{array}$	$CH_2=CHCH_3$
Butene	C_4H_8	$\begin{array}{cccc} H & H & H & H \\ C=C-C-C-H \\ H & H & H \end{array}$	$CH_2=CHCH_2CH_3$
Pentene	C_5H_{10}	$\begin{array}{ccccc} H & H & H & H & H \\ C=C-C-C-C-H \\ H & H & H & H \end{array}$	$CH_2=CHCH_2CH_2CH_3$
Hexene	C_6H_{12}	$\begin{array}{cccccc} H & H & H & H & H & H \\ C=C-C-C-C-C-H \\ H & H & H & H & H \end{array}$	$CH_2=CHCH_2CH_2CH_2CH_3$

Cracking

Crude oil contains a **mixture** of **hydrocarbons**. The large hydrocarbons separated during the **fractional distillation** of crude oil are not very useful. **Cracking** can break down these large hydrocarbons into **smaller, more useful** molecules. There is more demand and therefore a higher price for the smaller hydrocarbons like petrol compared with larger molecules like those in lubricating oil.

Industrial cracking

Cracking is an example of a thermal decomposition reaction. Large, less useful hydrocarbon molecules are broken down into smaller, more useful ones using heat and a hot aluminium oxide catalyst.

The ethene molecule is useful because it contains a double bond between two carbon atoms. Ethene can be used to make plastics and other useful substances.

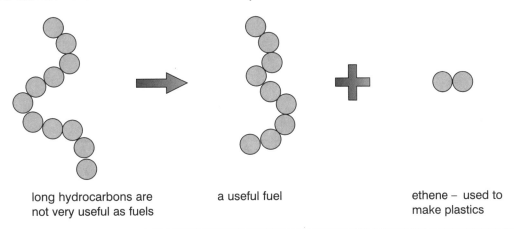

long hydrocarbons are not very useful as fuels

a useful fuel

ethene – used to make plastics

Quick Test

1. What is meant by an *unsaturated hydrocarbon*?

2. Name and draw the structural formula of the first member of the alkene family.

3. What is the general formula for the alkenes?

4. Write down the molecular formula and draw the structural formula for butene.

5. Why are long chain hydrocarbons cracked into smaller ones?

6. Which catalyst is used for cracking?

7. What is ethene used for?

Answers 1. A hydrocarbon which has a carbon to carbon double bond. **2.** Ethene **3.** C_nH_{2n} **4.** C_4H_8 **5.** To make smaller, more useful hydrocarbons. **6.** Aluminium oxide. **7.** Making plastics.

More about cracking

Cracking in the laboratory

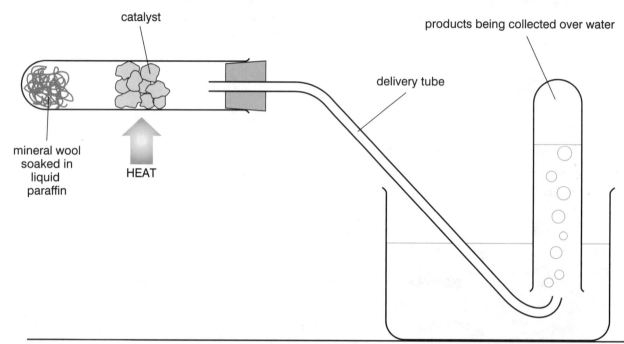

An experiment to demonstrate hydrocarbon cracking

The **catalyst** used in the laboratory can be either **aluminium oxide** or bits of **broken clay pots**. The long chain molecules in the liquid paraffin are broken into smaller molecules on the surface of the catalyst.

A mixture of gases is collected. The fact that the products are gases rather than liquids tells us that the product molecules are smaller than the reactant molecules. This means that **cracking** has taken place. The mixture of gases collected also decolourises bromine solution. This tells us that **unsaturated hydrocarbons** such as ethene are present in the products.

CREDIT

In industry the catalyst is used so that the reaction can take place at a lower temperature and so can be carried out at a lower cost. This industrial reaction is often known as catalytic cracking.

Why does cracking produce unsaturated hydrocarbons?

To answer this question we must first remember that when a molecule is cracked it breaks into smaller molecules, but it does not react with any other substances.

Consider cracking hexane as an example. Hexane has the molecular formula C_6H_{14}. This means that each molecule of hexane contains six carbon atoms and fourteen hydrogen atoms.

If hexane is cracked to produce propane, C_3H_8, what will be the other product? The other product must have three carbon atoms (C_6-C_3) and six hydrogen atoms ($H_{14}-H_8$) so will have the molecular formula C_3H_6 and so is propene.

If hexane is cracked to produce ethane, C_2H_6, what will be the other product?

The other product must have four carbon atoms (C_6-C_2) and eight hydrogen atoms ($H_{14}-H_6$) so will have molecular formula C_4H_8 and so is butene.

No matter which C–C bond in hexane is broken, one of the products must be an unsaturated alkene.

Quick Test

1. Which catalyst can be used in the laboratory for the cracking of liquid alkanes?

2. Draw a labelled diagram to show the apparatus that would be used in the laboratory to crack liquid paraffin.

3. Name and write the molecular formula for the other products when the following cracking reactions take place:
   a) heptane cracked to give butane
   b) pentane cracked to give propane
   c) hexane cracked to give methane.

Addition reactions

What is an addition reaction?

When a **diatomic molecule** adds on across the **C=C double bond** in an alkene it is known as an **addition reaction**.

The **reactant alkene** is unsaturated as it has a C=C double bond and the product is saturated since it has no C=C double bonds. When **hydrogen adds on** across the C=C double bond the **product is an alkane**.

Top Tip
This page is very important. Make sure you know about addition reactions and the test for unsaturated hydrocarbons.

```
  H H H                          H H H
  | | |                          | | |
  C=C-C-H   +   H-H    ⟶    H-C-C-C-H
  | |                            | | |
  H H                            H H H
 propene       hydrogen          propane
```

If hydrogen were to add on to butene the product would be butane.

If a halogen such as bromine adds on across the C=C double bond, the structural formula of the product must show the bromine atoms joined on to the carbon atoms where the double bond had been. This is shown in the examples below.

```
  H H H                          H H H
  | | |                          | | |
  C=C-C-H   +   Br-Br    ⟶    H-C-C-C-H
  | |                            | | |
  H H                           Br Br H
```

```
  H H H H                          H H H H
  | | | |                          | | | |
  C=C-C-C-H   +   Br-Br    ⟶    H-C-C-C-C-H
  | | |                            | | | |
  H H H                           Br Br H H
```

```
  H H H H                          H H H H
  | | | |                          | | | |
H-C-C=C-C-H   +   Br-Br    ⟶    H-C-C-C-C-H
  |     |                          | | | |
  H     H                          H Br Br H
```

Test for unsaturated hydrocarbons

When orange-brown **bromine solution** is reacted with an alkene the bromine **adds on** across the C=C double bond forming a colourless product. This reaction is used as a test for unsaturated hydrocarbons.

When bromine solution is added to an unsaturated hydrocarbon the bromine solution is decolourised rapidly:

$$
\underset{\text{ethene}}{\overset{\displaystyle \text{H\ \ H}}{\underset{\displaystyle \text{H\ \ H}}{\text{C=C}}}} \quad + \quad \underset{\text{bromine}}{\text{Br-Br}} \quad \longrightarrow \quad \underset{\text{colourless product}}{\overset{\displaystyle \text{H\ \ H}}{\underset{\displaystyle \text{Br\ Br}}{\text{H-C-C-H}}}}
$$

Bromine solution is not decolourised when it is added to a **saturated hydrocarbon** and so this is used as a **test for unsaturated hydrocarbons**.

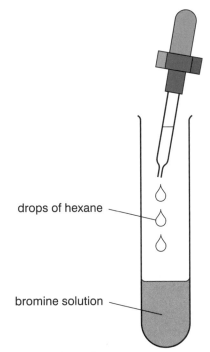

drops of hexane

bromine solution

no reaction with the saturated alkane.
The bromine solution stays orange-brown

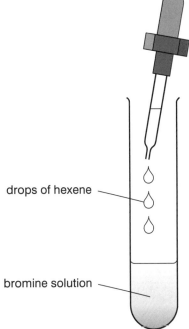

drops of hexene

bromine solution

addition reaction with the unsaturated alkene.
The bromine solution loses its colour rapidly

Quick Test

1. What is the test for *unsaturated hydrocarbons*?

2. What will be the product when hydrogen is added to ethene?

3. Why will butane not take part in addition reactions?

4. Using structural formulae, write an equation for the addition reaction between propene and bromine.

Answers 1. Unsaturated hydrocarbons decolourise bromine solution rapidly. **2.** Ethane. **3.** Butane does not have a C=C double bond.

4.

$$
\overset{\displaystyle \text{H\ \ H}}{\underset{\displaystyle \text{H\ H\ H}}{\text{C=C-C-H}}} + \text{Br}_2 \longrightarrow \overset{\displaystyle \text{H\ \ H}}{\underset{\displaystyle \text{Br\ Br\ H}}{\text{H-C-C-C-H}}}
$$

Cycloalkanes and isomers

The cycloalkane family

The **cycloalkanes** are another series of **hydrocarbons**. Cycloalkane molecules contain hydrogen and carbon only.

The carbon atoms in cycloalkanes **form a ring** rather than a chain. All the carbon-to-carbon bonds are **single bonds** so the cycloalkanes are **saturated** hydrocarbons. This means that cycloalkanes do not decolourise bromine solution rapidly.

Members of the cycloalkane family make up another **homologous series**. The general formula for the cycloalkane homologous series is C_nH_{2n}.

- The simplest cycloalkane is cyclopropane. It has molecular formula C_3H_6.
- Cyclobutane is the second member of the cycloalkane homologous series. It has molecular formula C_4H_8.

Formulae of the cycloalkanes

The names, molecular formulae and structural formulae of the first four members of the cycloalkane family are shown in the table below.

Name	Molecular formula	Structural formula
Cyclopropane	C_3H_6	
Cyclobutane	C_4H_8	
Cyclopentane	C_5H_{10}	
Cyclohexane	C_6H_{12}	

CREDIT

What are isomers?

Isomers have the same molecular formula but have different structural formulae. For example, butene and cyclobutane are isomers because they both have molecular formula C_4H_8 but their structures are different.

There are two isomers of butane, both with molecular formula C_4H_{10}.

Larger hydrocarbons have more isomers. Pentane has three isomers, hexane has five isomers and heptane has nine isomers.

Top Tip
Make sure you know what isomers are and that you can draw isomers of these hydrocarbons.

Quick Test

1. What is the molecular formula for cyclobutane?

2. What is the general formula for the cycloalkane family?

3. Draw the structural formula for cyclobutane.

4. Which two compounds have molecular formula C_3H_6?

5. What are isomers?

6. Name an isomer of cyclopentane.

7. Draw the structures of the three isomers of pentane.

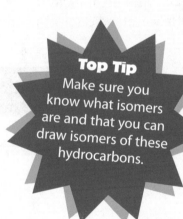

Answers 1. C_4H_8 **2.** C_nH_{2n} **3.** **4.** Propene and cyclopropane. **5.** Isomers have the same molecular formula but different structural formulae. **6.** Pentene. **7.**

Acids and alkalis

pH scale and indicators

The pH scale is a measure of the **concentration of hydrogen ions** (H^+) in a solution. The **more hydrogen ions** that are present, **the more acidic** the solution and **the lower the pH** value.

The pH scale is a continuous scale which runs from below 0 to above 14, but the pH values of most solutions you work with lie between 1 and 14.

- Acids have pH values below 7.
- Alkalis have pH values above 7
- Neutral solutions such as pure water have pH = 7.

Indicators are used to show whether a solution is acidic, alkaline or neutral. An indicator will be one colour in an acid and a different colour in an alkali. There are many different indicators but universal indicator is the one you will use most. Universal indicator is red in acid, green when it is neutral and blue/purple in an alkali.

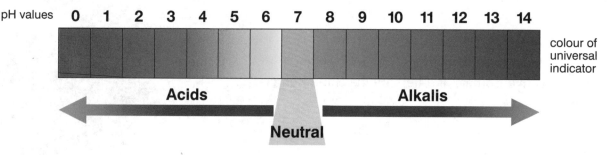

The pH scale

The pH of a substance can also be tested using moist pH paper. pH paper turns the same colours as universal indicator.

Diluting acids and alkalis

Adding water decreases the concentration of acids and alkalis.

- When **acids are diluted the pH increases**. Eventually the pH will increase to 7 as the solution becomes neutral. The pH will remain at 7 even if more water is added.

- When **alkalis are diluted the pH decreases**. Eventually the pH will decrease to 7 as the solution becomes neutral.

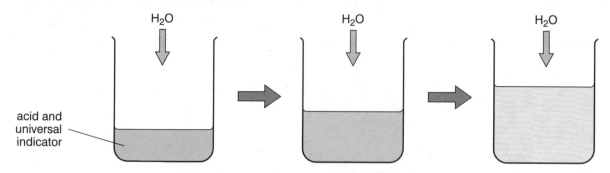

Water being added to a beaker of acid to neutralise it

Common acids and alkalis

Acids

- Acids are corrosive.
- Acidic solutions have a pH less than 7.

Common laboratory acids are:

- hydrochloric acid
- sulphuric acid
- nitric acid.

Common household acids are:

- vinegar
- lemonade
- fruit juices.

The soluble oxides of non-metals form acidic solutions. Examples are CO_2 which dissolves to give carbonic acid and SO_2 which dissolves to give sulphurous acid (acid rain). Nitrogen dioxide also dissolves to form an acid.

Alkalis

- Alkalis are corrosive.
- Alkalis are also called bases.
- Alkalis are soluble bases.
- Alkalis have a pH of more than 7.

Common laboratory alkalis are:

- sodium hydroxide
- potassium hydroxide
- calcium hydroxide (lime water)
- ammonia.

Common household alkalis are:

- baking soda solution
- oven cleaner.

The soluble oxides and hydroxides of metals form alkaline solutions.

Ions in acids and alkalis

- All acid solutions contain hydrogen ions, H^+.
- During electrolysis of any acid, the positive hydrogen ions travel to the negative electrode where they change into hydrogen gas.
- All alkalis contain hydroxide ions, OH^-.
- Water contains equal quantities of H^+ and OH^- ions but their concentration is very small in pure water.
- In water and in neutral solutions, there are equal numbers of H^+ and OH^- ions.
- In acids, there are more H^+ ions than OH^- ions.
- In alkalis, there are more OH^- than H^+ ions.
- Diluting acids and alkalis reduces the concentrations of H^+ and OH^- ions.

Top Tip
Remember, acids contain H^+ ions, whereas alkalis contain OH^- ions.

CREDIT

Quick Test

1. What is the pH of a neutral solution?
2. Which ions are present in all acids.
3. Write the formula for the ion which is present in all alkalis.

4. Name three laboratory acids.
5. Name two household alkalis
6. What colour is universal indicator in alkalis?

Answers 1. 7 **2.** Hydrogen ions, H^+ **3.** OH^- **4.** Hydrochloric, nitric and sulphuric acids. **5.** Baking soda and oven cleaner. **6.** Blue-purple.

Reactions of acids

Acids and metals

When an acid reacts with a metal such as magnesium, **a salt** and **hydrogen gas** are produced (this does not happen with all metals). The hydrogen produced will burn with a squeaky pop.

'squeaky' pop

H_2 gas

burning splint

Mg and H_2SO_4

Hydrogen is produced when an acid reacts with magnesium

Top Tip
Reactions of acids are very important. You should know the formulae for the different acids, the products of these reactions and be able to write formula equations for the reactions.

The word equation for the reaction is:

sulphuric acid + magnesium → magnesium sulphate + hydrogen

The formula equation is:

$H_2SO_4 + Mg \rightarrow MgSO_4 + H_2$

This reaction involves both oxidation and reduction and it is known as a **redox reaction**. The sulphate ions do not change in the reaction. Ions which do not change in a reaction are known as **spectator ions**.

Sulphuric acid contains hydrogen ions, H^+ and sulphate ions, SO_4^{2-}. In the reaction, magnesium metal is reacting with hydrogen ions from the acid. Magnesium atoms are losing electrons to the hydrogen ions. This means that the magnesium atoms are being **oxidised** and the hydrogen ions are being **reduced**.

 CREDIT

Acids and alkalis

Alkalis will **neutralise** acids forming a salt and water. The word equation for sodium hydroxide neutralising nitric acid is:

nitric acid + sodium hydroxide → sodium nitrate + water

The formula equation is:

$HNO_3 + NaOH \rightarrow NaNO_3 + H_2O$

 CREDIT

In the reaction, the H^+ ions from the acid are reacting with the OH^- ions from the alkali to form water:

$H^+ + OH^- \rightarrow H_2O$

The other ions present are spectator ions.

Acids and metal oxides

Metal oxides will **neutralise** acids to form a salt and water. The word equation for copper(II) oxide reacting with sulphuric acid is:

sulphuric acid + copper(II) oxide → copper(II) sulphate + water

The formula equation is: $H_2SO_4 + CuO \rightarrow CuSO_4 + H_2O$

In the reaction the H^+ ions from the acid are reacting with the O^{2-} ions from the metal oxide to form water:

$2H^+ + O^{2-} \rightarrow H_2O$

The other ions present are spectator ions.

CREDIT

Top Tip

When an acid reacts it is the hydrogen ions from the acid that change. The negative ion in an acid usually behaves as a spectator ion.

Acids and metal carbonates

Metal carbonates will **neutralise** acids to form a salt, water and carbon dioxide. In the reaction, bubbles of carbon dioxide gas will be produced. The carbon dioxide formed turns lime water cloudy.

The word equation for hydrochloric acid reacting with magnesium carbonate is:

hydrochloric acid + magnesium carbonate → magnesium chloride + water + carbon dioxide

The formula equation is:

$HCl + MgCO_3 \rightarrow MgCl_2 + H_2O + CO_2$

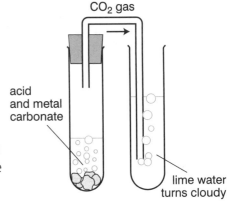

Carbon dioxide is produced when a metal carbonate reacts with acid

In the reaction the H^+ ions from the acid are reacting with the CO_3^{2-} ions from the metal carbonate to form water and carbon dioxide:

$2H^+ + CO_3^{2-} \rightarrow H_2O + CO_2$

The other ions present are spectator ions.

CREDIT

Quick Test

1. Which gas is produced when magnesium reacts with hydrochloric acid?

2. Write the balanced formula equation for the reaction of magnesium with hydrochloric acid.

3. Which ion is present in all acid solutions?

4. What are spectator ions?

5. What are the two products when an acid is neutralised by an alkali?

6. What is the test for the gas produced when an acid is neutralised by a metal carbonate?

7. Write down three types of compounds that will neutralise acids.

8. What is the formula for
   a) the oxide ion
   b) the carbonate ion?

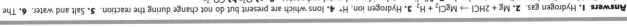

Answers 1. Hydrogen gas. **2.** $Mg + 2HCl \rightarrow MgCl_2 + H_2$ **3.** Hydrogen ion, H^+ **4.** Ions which are present but do not change during the reaction. **5.** Salt and water. **6.** The carbon dioxide produced turns lime water cloudy. **7.** Alkalis, metal oxides and metal carbonates. **8. a)** O^{2-} **b)** CO_3^{2-}

Neutralisation

What is neutralisation?

- A **neutralisation reaction** is one in which a substance cancels out the effect of an acid.
- A substance which neutralises an acid is called a **neutraliser**.
- In a neutralisation reaction the **pH of an acid rises towards 7**.
- In a neutralisation reaction the **pH of an alkali decreases towards 7**.
- **Bases** are neutralisers.
- A base is the **chemical opposite of an acid**.
- Examples of bases are **alkalis**, **metal oxides** and **metal carbonates**.
- **Ammonia** is another example of **a base**.
- During a neutralisation reaction, **water and a salt** are always formed.

General equations for neutralisation reactions

When an **acid** reacts with an **alkali** the general equation is:

acid + alkali → salt + water

When an **acid** reacts with a **metal oxide** the general equation is:

acid + metal oxide → salt + water

When an **acid** reacts with a **metal carbonate** the general equation is:

acid + metal carbonate → salt + water + carbon dioxide

In all these reactions of acids, it is the **hydrogen ion, H^+**, from the acid that is reacting

Top Tip
Note that salts are produced in neutralisation reactions.

Everyday examples of neutralisation

- Putting vinegar onto the alkaline sting from a wasp.
- Putting baking soda solution onto the acid sting from a bee.
- Spreading lime on fields in which the soil is too acidic.
- Taking antacid tablets to treat acid indigestion in your stomach.
- Using lime (or limestone) to reduce the effects of acid rain on lakes.

Vinegar can be used to treat a wasp sting

Names of salts produced

When an acid reacts with alkalis, metal oxides, metal carbonates or metals a salt is produced.

The name of the salt produced depends on the acid used and the neutraliser it reacts with.

Neutralising **hydrochloric** acid will produce **chloride** salts:

- **sodium** hydroxide + hydro**chloric** acid → **sodium chloride** + water

Neutralising **nitric** acid will produce nitrate salts:

- **calcium** hydroxide + **nitric** acid → **calcium nitrate** + water

Neutralising sulphuric acid will produce sulphate salts:

- **potassium** hydroxide + **sulphuric** acid → **potassium sulphate** + water

Quick Test

1. What is a *base*?
2. Write down three examples of bases.
3. What happens to the pH of an acid during a neutralisation reaction?
4. What should you apply to a wasp sting to relieve the pain?
5. Write the general equation for an acid reacting with an alkali.
6. Write the general equation for an acid reacting with a metal oxide.
7. Write the general equation for an acid reacting with a metal carbonate.
8. What products are always formed in a neutralisation reaction?
9. Which acid produces nitrate salts when it is neutralised?
10. Write the chemical formula for the acid which produces sulphate salts when it is neutralised.
11. Write the chemical equation for hydrochloric acid reacting with sodium hydroxide.
12. What substance is put on soils which are too acidic and is also sprayed onto lakes which have been affected by acid rain?

Answers 1. A base is the opposite of an acid and will neutralise acids. **2.** Examples of bases include alkalis, metal oxides and metal carbonates. **3.** It rises towards 7. **4.** Vinegar or any very dilute acid. **5.** acid + alkali → salt + water **6.** acid + metal oxide → salt + water **7.** acid + metal carbonate → salt + water + carbon dioxide **8.** A salt and water. **9.** Nitric acid. **10.** H_2SO_4. **11.** HCl + NaOH → NaCl + H_2O **12.** Lime.

Preparation of salts

Insoluble salts

When asked to prepare a salt, it is necessary to find out first if the salt is **soluble** or **insoluble**. This can be done by looking at page 5 of the Data Booklet.

If the salt to be prepared is **insoluble**, then it must be prepared by **precipitation**. This is done by preparing one solution containing the positive ions in the salt and another solution containing the negative ions in the salt. The two solutions are then mixed together to form a precipitate of the salt which is then filtered off.

For example, insoluble lead iodide is made by preparing a solution containing lead ions, such as **lead** nitrate, and another solution containing iodide ions, such as potassium **iodide**. When these two solutions are mixed together a precipitate of **lead iodide** immediately forms:

$$Pb(NO_3)_2(aq) + 2KI(aq) \rightarrow PbI_2(s) + 2KNO_3(aq)$$

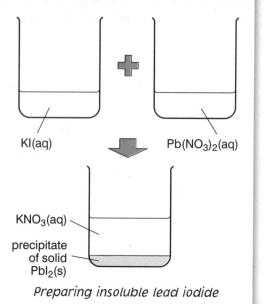

KI(aq)   Pb(NO$_3$)$_2$(aq)

KNO$_3$(aq)

precipitate of solid PbI$_2$(s)

Preparing insoluble lead iodide

Soluble salts

Soluble salts are prepared from the appropriate acid.

Acid and alkali

The acid and alkali are mixed together until the solution is neutral. This can be checked with pH paper. The water in the salt solution formed is then evaporated off leaving the solid salt behind. This is a good method of preparing sodium and potassium salts.

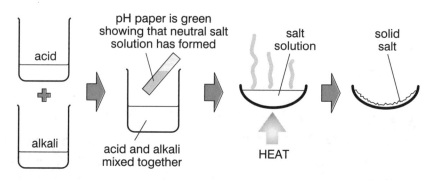

acid

alkali

pH paper is green showing that neutral salt solution has formed

acid and alkali mixed together

salt solution

HEAT

solid salt

Preparing a soluble salt

Other methods involve reacting the appropriate acid with a solid such as a metal, metal oxide or metal carbonate. The general method is to add excess of the solid to make sure that all the acid has been used up. The solid left over is then filtered off and the salt solution passes through the filter paper. The solid salt is then isolated by evaporating off the water

Acid and metal

Metals can be reacted with acids to form a salt and hydrogen:

metal + acid → salt + hydrogen

zinc + hydrochloric acid → zinc chloride + hydrogen

$$Zn(s) + 2HCl(aq) \rightarrow ZnCl_2(aq) + H_2(g)$$

magnesium + sulphuric acid → magnesium sulphate + hydrogen

$$Mg(s) + H_2SO_4(aq) \rightarrow MgSO_4(aq) \quad + \quad H_2(g)$$

Top Tip
Remember, the solid has to be added in excess to make sure the acid is used up. The excess solid is filtered and the salt solution is heated to evaporate off the water.

Acid and metal oxide

Metal oxides are also bases; they can be reacted with acids to make salts and water:

metal oxide + acid → salt + water

copper oxide + hydrochloric acid → copper chloride + water

$CuO(s)$ + $2HCl(aq)$ → $CuCl_2(aq)$ + $H_2O(l)$

zinc oxide + sulphuric acid → zinc sulphate + water

$ZnO(s)$ + $H_2SO_4(aq)$ → $ZnSO_4(aq)$ + $H_2O(l)$

Top Tip
Sulphuric acid makes sulphate salts. Hydrochloric acid makes chloride salts. Nitric acid makes nitrate salts.

Acid and metal carbonate

Metal carbonates are also bases; they can be reacted with acids to make salts.

Copper chloride can be made using hydrochloric acid and copper carbonate:

copper carbonate + hydrochloric acid → copper chloride + water + carbon dioxide

$CuCO_3(s)$ + $2HCl(aq)$ → $CuCl_2(aq)$ + $H_2O(l)$ + $CO_2(g)$

- Copper carbonate is added to the acid until it stops fizzing.
- The unreacted copper carbonate is then removed by **filtering**.
- The solution is poured into an evaporating dish.
- It is heated until the first crystals appear.
- The solution is then left for a few days for the copper chloride to **crystallize**.

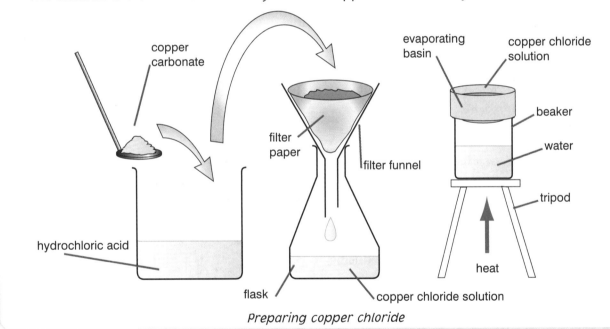

Preparing copper chloride

Quick Test

1. What type of reaction is used to prepare insoluble salts?

2. Which method would be used to prepare sodium nitrate?

3. Which acid and metal would be used to prepare zinc chloride?

4. Which acid and metal oxide would be used to prepare copper sulphate?

5. Which acid and metal carbonate would be used to prepare calcium nitrate?

6. Which method would be used to prepare silver chloride?

Answers 1. Precipitation. 2. Acid and alkali. 3. Hydrochloric acid and zinc. 4. Sulphuric acid and copper oxide. 5. Nitric acid and calcium carbonate. 6. Precipitation.

Electricity from chemicals

Batteries or cells

The words **battery** or **cell** are often used as if they were the same thing, however, a battery is a collection of cells joined together. When a battery is working there is an energy change taking place. **Chemical energy** inside the battery is changing in to **electrical energy**. The electricity coming from the battery is made as a result of a **chemical reaction**. The electricity can then flow through a circuit lighting a bulb or turning a motor, for example. Electricity passing along metal wires is a **flow of electrons**.

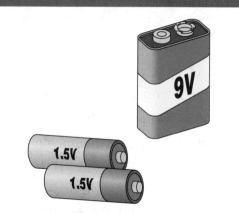

Various batteries and cells

Like other chemical reactions, **the reaction slows down** as the **chemicals are used up**, and the battery will no longer work once the chemicals have been used up. At this stage the battery should be disposed of carefully unless it is rechargeable.

An advantage that batteries have over mains electricity is that they produce **lower voltages** and so are safer to use. They also allow appliances to be **portable** so they can be used where there is no mains electricity.

Top Tip
Remember, electricity from a battery is produced from a chemical reaction.

Lead-acid batteries

Lead-acid batteries contain lead plates dipped into sulphuric acid. Car batteries are lead-acid.

When a lead-acid battery is no longer producing electricity it can be recharged using a battery charger connected to mains electricity.

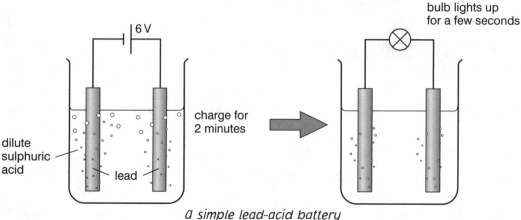

6 V

dilute sulphuric acid

lead

charge for 2 minutes

bulb lights up for a few seconds

a simple lead-acid battery

Simple electrochemical cells

The simplest electrochemical cell consists of two different metals, such as zinc and copper dipping into an **electrolyte**. If this is connected to a voltmeter, there will be a voltage reading and the direction of electron flow is from zinc to copper.

If there is no electrolyte present then there will be no reading on the voltmeter. The purpose of the electrolyte is to complete the circuit, allowing electrons to flow through the wires and ions to flow through the solution.

Remember, an electrolyte is a solution containing ions.

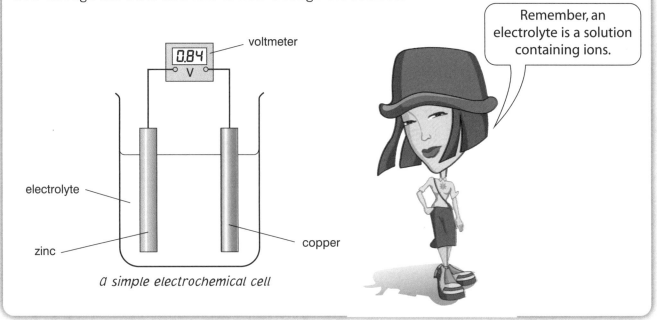

voltmeter

electrolyte

zinc

copper

a simple electrochemical cell

Quick Test

1. What energy change takes place inside a battery when it is operating?

2. What produces the electricity from a cell or a battery?

3. Why does a battery eventually stop working?

4. Give two advantages that batteries have over mains electricity.

5. How are lead-acid batteries recharged and what are they used for?

6. What is an electrolyte?

7. What does a simple electrochemical cell contain?

8. What is the purpose of an electrolyte in an electrochemical cell?

Answers 1. Chemical to electrical. **2.** It is the result of a chemical reaction. **3.** One or more of the chemicals gets used up. **4.** Batteries are safer as they produce a lower current and lower voltage. Batteries can also be used for portable appliances. **5.** A battery charger connected to the mains electricity supply recharges them. They are used in cars. **6.** An ionic solution. **7.** Two different metals dipping into an electrolyte. **8.** To complete the circuit.

Using the electrochemical series

The electrochemical series

When different metals are connected together, as shown below, an electric current flows from one metal to the other.

This leads to the **electrochemical series** (e.c.s.) which is given on page 7 the Data Booklet. It is a list of mainly metals, with the most reactive metals at the top and the least reactive metals at the bottom. If the metals are connected together, as shown on the right, then **electrons will flow from the metal higher in the electrochemical series to the metal lower in the series**.

The further apart the metals are in the electrochemical series then the greater the voltage and current produced.

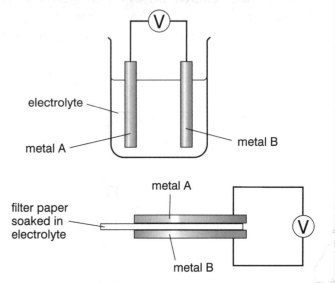

an electric current flowing from one metal to another

More complicated cells

Electricity can be produced in a cell by connecting two different metals in solutions of their metal ions as shown in the cells below.

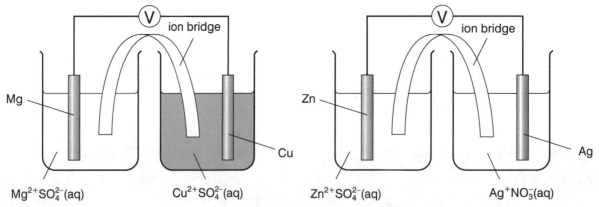

$Mg^{2+}SO_4^{2-}(aq)$ $Cu^{2+}SO_4^{2-}(aq)$ $Zn^{2+}SO_4^{2-}(aq)$ $Ag^{+}NO_3^{-}(aq)$

Electricity being produced by connecting two different metals in solutions of their metal ions

Because magnesium is higher than copper in the electrochemical series, and zinc is higher than silver, we can predict correctly that the electrons will flow from magnesium to copper and from zinc to silver in the above cells.

The purpose of the '**ion bridge**' is to complete the circuit. The ion bridge is often a piece of filter paper soaked in an electrolyte such as salt solution.

The ion bridge completes the circuit by allowing ions to move through it and pass from beaker to beaker. This allows electrons to move through the wires.

Cells without metals

Looking at the electrochemical series on page 7 of the Data Booklet you can see that there are equations which do not have metals in them. This means that not all cells need to contain metals. A carbon rod can be used as the electrode as it is cheap and allows electrons to flow through it.

Top Tip

Remember, electrons flow from the material higher in the electrochemical series to the material lower down in the series.

CREDIT

Electrochemical cells without metals using carbon rods as the electrodes

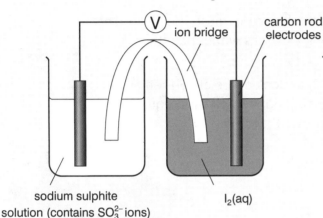

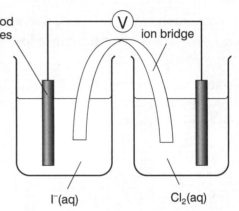

sodium sulphite solution (contains SO_3^{2-} ions)

I_2(aq)

I^-(aq)

Cl_2(aq)

In this cell, electrons flow from left to right since SO_3^{2-} is above I_2 in the electrochemical series. Note that the sodium ions are very unreactive and behave as spectator ions.

In this cell electrons flow from left to right since I^- is above Cl_2 in the electrochemical series.

Quick Test

1. In which direction will electrons flow in an electrochemical cell containing magnesium connected to copper?

2. Which will produce the larger voltage in an electrochemical cell, magnesium connected to silver or zinc connected to copper?

3. What is the purpose of the ion bridge in a cell?

4. In which direction will the electrons flow in a cell containing chlorine (Cl_2) solution and iodide (I^-) ions?

5. Give a reason for your answer to question 4.

6. Why are sodium ions often spectator ions in chemical reactions?

Answers 1. From magnesium to copper. **2.** Magnesium connected to silver. **3.** To complete the circuit. **4.** From I^- to Cl_2. **5.** Because iodine is higher than chlorine in the electrochemical series. **6.** Because sodium ions are very unreactive.

Oxidation, reduction and displacement

Oxidation and reduction

- When a metal or any other substance is giving electrons away in a cell, this is an example of oxidation.

- When a substance in a cell is gaining electrons this is an example of reduction.

- When an oxidation reaction occurs, a reduction reaction must also occur. This is because you cannot get a substance losing electrons without another substance gaining these electrons.

The ion-electron equations on page 7 of the Data Booklet are all written as reduction ion-electron equations. To write an oxidation ion-electron equation, first find the relevant equation in the Data Booklet then reverse it.

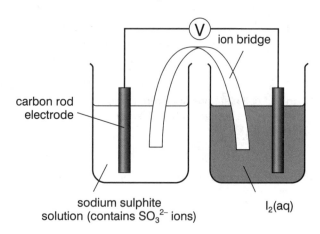

carbon rod electrode

sodium sulphite solution (contains SO_3^{2-} ions)

I_2(aq)

> **Top Tip**
> Remember, oxidation is the loss of electrons, whereas reduction is the gain of electrons.

For example, for this cell, the two relevant ion electron equations from the Data Booklet are:

$$SO_4^{2-} + 2H^+ + 2e^- \rightarrow SO_3^{2-} + H_2O \text{ and}$$

$$I_2 + 2e^- \rightarrow 2I^-$$

These are both reduction equations. The upper one has to be reversed and so it becomes oxidation:

$$SO_3^{2-} + H_2O \rightarrow SO_4^{2-} + 2H+ + 2e^-$$

When the oxidation and reduction ion-electron equations are combined together we get the overall **redox equation**.

> **Top Tip**
> Remember, when you find the two ion-electron equations in the Data Booklet, the upper one has to be reversed, the lower one stays unchanged.

Displacement reactions

The higher a metal is in the electrochemical series, the more easily it will lose electrons to form ions and the more difficult it will be to reduce these ions back to the metal atoms. The ions of the metals nearer the bottom of the electrochemical series are more likely to accept electrons to turn back into metal atoms.

As a result, if a metal is added to a solution containing ions of a metal lower in the electrochemical series, a displacement reaction takes place. For example, if magnesium is put into copper sulphate solution the magnesium will displace copper. This is because magnesium is higher than copper in the electrochemical series.

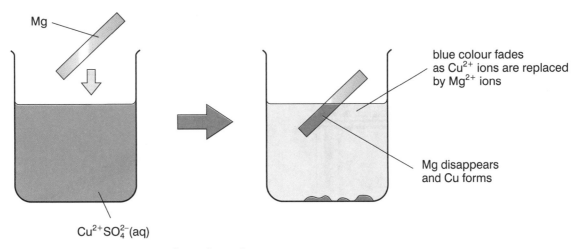

Mg

blue colour fades as Cu^{2+} ions are replaced by Mg^{2+} ions

Mg disappears and Cu forms

$Cu^{2+}SO_4^{2-}(aq)$

Magnesium displacing copper

$$Mg(s) + Cu^{2+}SO_4^{2-}(aq) \rightarrow Mg^{2+}SO_4^{2-}(aq) + Cu(s)$$

The electrochemical series can be used to predict whether a displacement reaction will take place. The metal higher in the electrochemical series will displace a metal lower down.

Quick Test

1. Write the ion-electron equation for the reduction of chlorine, Cl_2.

2. Explain why Cu will not displace Zn from $ZnSO_4$ solution.

Answers 1. $Cl_2 + 2e^- \rightarrow 2Cl^-$ **2.** Cu is below Zn in the electrochemical series.

Metals

Metals in the Periodic Table

Approximately three-quarters of the elements in the Periodic Table are **metals**. The metals are to be found at the left hand side of the zig-zag line shown in the Periodic Table on page 8 of your Data Booklet.

The uses to which we put metals are related to their distinctive properties.

Properties of metals

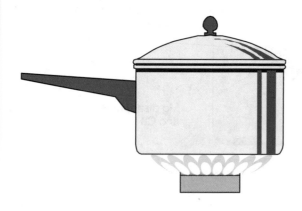

Metals are good conductors of heat.

Metals have high **melting** and **boiling points**. All the metals are solids except mercury, which is **liquid** at room temperature.

Metals are good conductors of electricity.

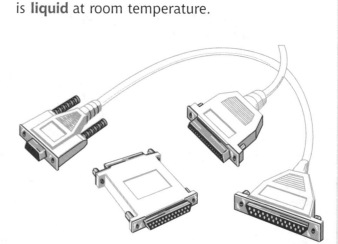

Metals are **strong** and **dense**, but they are also **malleable** (can be hammered into shape) and **ductile** (can be drawn into wires).

Top Tip
You should know the properties of metals, especially that all metals are good conductors when solid or liquid.

Recycling metals

Metals are very useful but will not last forever. Copper, for example, is very useful but is in very short supply. We need to use metals more carefully and avoid wasting them. We can all help to make metals last longer by recycling scrap metals so that they can be used again. Recycling also saves money and stops used metals causing a litter problem.

Alloys

Top Tip
Alloys are two or more metals mixed together. You should know which metals are in at least one alloy and a use for that alloy.

The properties of metals can be improved by making them into **alloys**. An alloy is formed when **two or more metals are mixed together**, or, in some cases, when a metal is mixed with a non-metal. Alloys are usually made by melting the mixture and then allowing it to cool until it has **solidified**. Common alloys and their uses are shown in the table below.

Name of alloy	Metals present	Uses
Brass	Copper and zinc	Door handles and fittings
Bronze	Copper and tin	Statues

Name of alloy	Metals present	Uses
Solder	Tin and lead	Joining metals together
Stainless steel	Iron, chromium and nickel	Cutlery, kitchen sinks

Quick Test

1. Where are metals in the Periodic Table?
2. Approximately what fraction of the elements are metals?
3. Name four properties that are common to most metals.
4. Which metal is not solid at room temperature?
5. What is an alloy?
6. Why are alloys made?
7. Name two alloys and give a use for each.

Answers 1. They are at the left hand side of the zig-zag line. **2.** Approximately three-quarters (about 75 per cent). **3.** Good conductors of heat and electricity, high melting and boiling points, strong/dense, malleable and ductile. **4.** Mercury; it is the only liquid metal at room temperature. **5.** A mixture of metals. **6.** To improve the properties of metals. **7.** Any two from: brass (door handles and fittings), solder (joining metals), bronze (making statues), stainless steel (cutlery).

More about metals

Uses of some metals

Iron, nickel and copper are examples of transition metals. Some uses of these three important metals are given below.

Copper

- Copper is a **good conductor** of heat and electricity. It can be bent easily and does not **corrode**.
- Copper is used for electrical wiring because it can be bent into shape and is a good conductor of electricity.
- Copper is also used for making water pipes because it does not corrode and can be bent into shape without fracturing.

Iron

- Iron is **strong** but **brittle**.
- Iron is often made into **steel**.
- Steel is strong and cheap and is used in vast quantities; unfortunately it is also **heavy** and may **rust**.
- Iron and steel are useful **structural** materials. Bridges, buildings, ships, cars and trains are all constructed from these materials.
- Iron is a useful **catalyst** and is used in the Haber process.
- Stainless steel does not rust, but is more expensive to produce than iron.

Nickel

- Nickel is **hard**, **shiny** and **dense**.
- It is used to make **coins**.
- Nickel is also used as a catalyst in the manufacture of margarine.

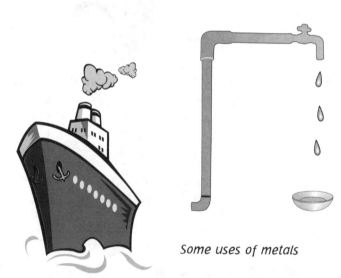

Some uses of metals

Top Tip
Metals usually have high melting and boiling points. An exception is mercury which is liquid at room temperature.

Bonding in metals

Metals have a **giant** structure. The electrons in the highest energy shells (outer electrons) are free to move through the whole structure.

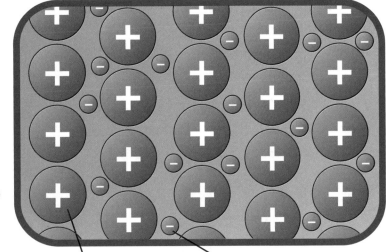

Metals have a giant structure. The electrons in the highest energy shells (outer electrons) are free to move through the whole structure

positive metal ions   sea of negative electrons

- The **free electrons** hold the ions together in a regular structure and give metals their special properties.
- The free electrons allow metals to **conduct** heat and electricity.
- The free electrons also allow the atoms to slide over each other, without the metal breaking.
- This means that metals can be drawn into wires and hammered into shape.

Top Tip
Metals conduct electricity because their outer electrons are free to move.

Quick Test

1. Write down two uses for copper.

2. Write down two uses for iron.

3. Write down two uses for nickel.

4. Describe, briefly, the structure of metals.

5. How do metals conduct electricity?

Answers 1. Electrical wires; water pipes. **2.** Catalyst in the Haber process; building cars, ships, trains, etc. **3.** Catalyst in the manufacture of margarine; making coins. **4.** Metals are composed of positive metal ions in a regular structure surrounded by a moving sea of electrons. **5.** The outer electrons are free to move and they carry the current.

Reactivity of metals

Reactivity series

Some metals are more reactive than others. The **reactivity series** places metals in order of reactivity.

Most reactive

potassium K

sodium Na

calcium Ca

magnesium Mg

zinc Zn

iron Fe

lead Pb

hydrogen H

copper Cu

silver Ag

Least reactive   gold Au

This order has been worked out by observing how vigorous the reaction is between the metal and:

- oxygen
- water
- dilute acid.

Top Tip
The reactivity series is very important. Make sure you know it. The order is very similar to the electrochemical series on page 7 of the Data Booklet.

Reacting metals with oxygen

When metals are heated with air they may react with the oxygen present.

metal   + oxygen → metal oxide

magnesium + oxygen → magnesium oxide

$2Mg(s)$ $+ O_2(g)$ $\rightarrow 2MgO(s)$

Most reactive

potassium K

sodium Na

calcium Ca   These metals react vigorously. They **burn** fiercely.

magnesium Mg

zinc Zn

iron Fe

lead Pb   These metals react **slowly** with air.

hydrogen H

copper Cu

silver Ag   No reaction

Least reactive   gold Au

Reacting metals with water

Some metals react with water to produce a metal hydroxide and hydrogen.

$$\text{metal} + \text{water} \rightarrow \text{metal hydroxide} + \text{hydrogen}$$
$$\text{sodium} + \text{water} \rightarrow \text{sodium hydroxide} + \text{hydrogen}$$
$$2Na(s) + 2H_2O(l) \rightarrow 2NaOH(aq) + H_2(g)$$

Most reactive

potassium K
sodium Na
calcium Ca
magnesium Mg

React **vigorously** with **cold water**

zinc Zn
iron Fe
lead Pb

React very **slowly** with water

hydrogen H

copper Cu
silver Ag
gold Au

No reaction

Least reactive

Top Tip
Remember that potassium and sodium react so violently with water that they must be stored in oil.

Reacting metals with dilute acids

Some metals (those more reactive than hydrogen) react with dilute acids to produce salts and hydrogen.

$$\text{metal} + \text{acid} \rightarrow \text{salt} + \text{hydrogen}$$
$$\text{calcium} + \text{hydrochloric acid} \rightarrow \text{calcium chloride} + \text{hydrogen}$$
$$Ca(s) + 2HCl(aq) \rightarrow CaCl_2(aq) + H_2(g)$$

Most reactive

potassium K
sodium Na
calcium Ca
magnesium Mg

React **violently** with dilute acid

zinc Zn
iron Fe
lead Pb

Reacts **slowly** with dilute acid

hydrogen H

copper Cu
silver Ag
gold Au

No reaction

Least reactive

magnesium   zinc   copper
Putting magnesium, zinc and copper in dilute acid

Quick Test

1. Which gas do metals react with when they burn in air?

2. What is formed when magnesium burns in air?

3. Write the balanced formula equation for magnesium burning in air.

4. Why does copper not react with hydrochloric acid?

5. Write the balanced formula equation for the reaction between magnesium and hydrochloric acid.

Answers 1. oxygen. 2. Magnesium oxide. 3. $2Mg + O_2 \rightarrow 2MgO$ 4. Copper is below hydrogen in the reactivity series. 5. $Mg + 2HCl \rightarrow MgCl_2 + H_2$

Other reactions of metals

Metal displacement reactions

A more reactive metal will displace a less reactive metal from a compound.

Most reactive

potassium K
sodium Na
calcium Ca
magnesium Mg
zinc Zn
iron Fe
lead Pb
hydrogen H
copper Cu
silver Ag
gold Au

Least reactive

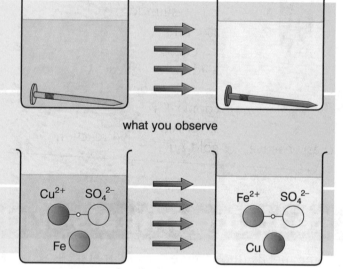

what you observe

what the ions are doing

Iron displacing copper from copper sulphate

- Iron is **more reactive** than copper.
- When an iron nail is placed in a solution of copper sulphate, the nail changes colour from silver to orange-pink.
- The solution changes colour from blue to a very pale green.
- This is an example of a displacement reaction. The more reactive metal, iron, **displaces the less reactive** metal, copper, from its compound, copper sulphate.

  iron + copper sulphate → copper + iron sulphate
  $Fe(s) + CuSO_4(aq) \rightarrow Cu(s) + FeSO_4(aq)$

- If the metal which is added is less reactive than the metal in the compound then no reaction will occur.

  copper + magnesium sulphate → no reaction

Top Tip

A metal reacting with an acid is similar to a displacement reaction.
- Metals above hydrogen in the reactivity series will displace hydrogen from an acid.
- Metals below hydrogen in the reactivity series will not react with acids to give hydrogen gas.

Corrosion of metals

When metals corrode, they lose electrons to form positive ions. This means that corrosion of metals is an oxidation process.

Group 1 metals

Group 1 metals corrode very quickly when exposed to air. They are stored in oil to stop air and water attacking them. If a piece of sodium is sliced to expose a shiny surface then the shine disappears rapidly as the sodium starts to corrode.

Gold and other precious metals

Gold and other precious metals corrode very slowly. Gold is so unreactive that a piece of gold jewellery will last for years and show no sign of corroding.

Iron

Iron **corrodes**, or **rusts**, faster than most transition metals. If either oxygen or water is completely removed then iron will **not** rust.

Top Tip
All metals corrode. The most reactive metals corrode very quickly. The least reactive metals corrode very slowly. Iron is the only metal which corrodes to form rust.

To preventing rusting the following measures can be used:

- **Coating the iron:** painting or coating iron in plastic or oil can stop oxygen and water from reaching it, but if the coating is damaged the iron will rust.

- **Alloying the metal:** if iron is mixed with other metals, such as chromium, it will form the alloy stainless steel. This does not rust.

- **Sacrificial protection:** if a metal which is more reactive than iron, such as zinc or magnesium, is connected to the iron, corrosion will be prevented. Because the zinc is more reactive, the zinc reacts instead of the iron. The iron is protected at the expense of the more reactive metal. For this reason it is called sacrificial protection. Speed-boat engines are protected by attaching a more reactive metal to them.

Quick Test

1. What is the rule for displacement reactions?

2. What is the word equation for the reaction between magnesium and copper sulphate?

3. Write a balanced symbol equation for the reaction.

4. What is the word equation for the reaction between zinc and iron sulphate?

5. Write a balanced symbol equation for the reaction.

6. What two things are needed for iron to rust?

7. What can be used to coat iron and prevent rusting?

8. How is stainless steel made and what are its advantages?

9. What is sacrificial protection?

Answers 1. A more reactive metal will displace a less reactive metal from a compound. **2.** Magnesium + copper sulphate → magnesium sulphate + copper **3.** $Mg(s) + CuSO_4(aq) \rightarrow MgSO_4(aq) + Cu(s)$ **4.** Zinc + iron sulphate → zinc sulphate + iron **5.** $Zn(s) + FeSO_4(aq) \rightarrow ZnSO_4(aq) + Fe(s)$ **6.** Oxygen + water **7.** Paint/plastic/oil **8.** By alloying iron with another metal such as Chromium, it does not rust. **9.** When a more reactive metal e.g. Zinc or Magnesium is put in contact with a less reactive metal e.g. Iron, the less reactive metal is protected as the Zn/Mg reacts first.

Corrosion in more detail

Corrosion is a chemical reaction in which the surface of the metal changes from an element into a compound.

Experiments on corrosion

Three nails were put into different test tubes as shown below and left for a few days.

- Test tube A contained a nail in dry air. The test tube also contained beads of a drying agent which remove water from the test tube.
- Test tube B contained a nail in water and air.
- Test tube C contained a nail in water only. Dissolved oxygen was removed from the water by boiling the water for about 20 minutes. There was also a layer of oil on the boiled water to prevent oxygen gas dissolving in the water.

Experiment to determine the conditions needed for rusting to take place

In test tubes A and C, the nail showed no sign of corroding. In test tube B, the nail rusted. This experiment shows that water and air (or oxygen) are needed for rusting to take place.

The experiment was repeated.

- This time test tube A contained a nail in salt water.
- Test tube B contained a nail in tap water.
- Test tube C contained a nail in acid rain.

Experiment to determine the conditions that speed up rusting

Top Tip
- Oxygen and water are needed for iron to rust.
- Electrolytes such as acid rain speed up rusting.

The nails rusted in all three test tubes. More rusting occurred in test tubes A and C. This experiment shows that salt and acid rain both speed up rusting. Salt solution and acids are electrolytes. Electrolytes speed up rusting and this is why salt spread on the roads in winter increases the rate of corrosion on cars.

Ferroxyl indicator

When iron begins to corrode it is oxidised to Fe^{2+} ions. Ferroxyl indicator turns a dark blue colour in the presence of Fe^{2+}, so it can be used as a test for Fe^{2+} ions.

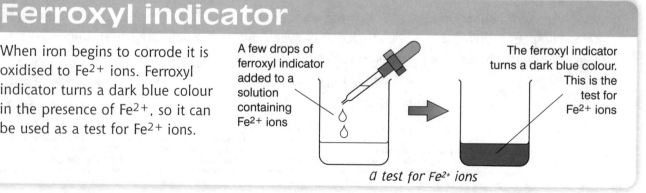

A few drops of ferroxyl indicator added to a solution containing Fe^{2+} ions

The ferroxyl indicator turns a dark blue colour. This is the test for Fe^{2+} ions

a test for Fe^{2+} ions

Equations on corrosion

Corrosion is an oxidation process and the first stage of iron rusting is when iron atoms lose electrons to form Fe^{2+} ions:

$$Fe \rightarrow Fe^{2+} + 2e^-$$

Molecules of water and oxygen pick up the electrons lost by the atoms of iron to form hydroxide ions:

$$2H_2O + O_2 + 4e^- \rightarrow 4OH^-$$

If there is an electrolyte present (such as acid or salt solution) then this will speed up the transfer of electrons.

The Fe^{2+} ions formed during corrosion lose another electron to form the brown coloured Fe^{3+} ions that are present in rust:

$$Fe^{2+} \rightarrow Fe^{3+} + e^-$$

CREDIT

Top Tip
Remember that ion-electron equations are given on page 7 of the Data Booklet.

Ways of preventing rusting

Stopping air and water getting to iron will stop it rusting. This can be done by coating the iron with **oil, grease, paint or plastic**. Iron can also be **electroplated** with another metal to prevent air and water getting to the iron.

Galvanised iron is iron coated in zinc. The zinc stops air and water getting to the iron. If the zinc gets scratched it will pass on electrons to the iron as zinc is higher in the electrochemical series. These electrons take the place of any electrons that the iron loses during corrosion and so stop the iron rusting. The zinc is therefore sacrificed to protect the iron and so this method is called **sacrificial protection.**

Any metal higher in the electrochemical series will pass on electrons to iron and protect it by sacrificial protection.

Another method of passing on electrons to iron is by connecting it to the **negative pole** of a **battery** or generator. Electrons will replace any being lost by the iron atoms as the iron starts to corrode. The negative pole of a car battery is connected to the car body to slow down the rate of corrosion of the car.

Two nails connected to 6 volts d.c. in salt solution and ferroxyl indicator

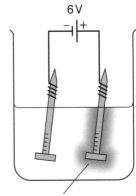

6 V

Dark blue colour formed around the nail connected to the positive pole but no corrosion around the nail connected to the negative pole

Quick Test

1. Which two substances are needed for iron to rust?

2. Which two substances speed up rusting?

3. Which indicator is used to show that iron is beginning to corrode? What colour does it turn?

4. Write the ion-electron equation for iron starting to corrode.

5. Give two ways of passing on electrons to iron to prevent it rusting.

Answers 1. Oxygen (air) and water. **2.** Salt solution and acid rain. **3.** Ferroxyl indicator turns dark blue. **4.** $Fe \rightarrow Fe^{2+} + 2e^-$ **5.** Connecting iron to a metal higher in the e.c.s. or to the negative pole of a battery or generator.

Extraction of metals

Metal ores

Only the very unreactive metals such as silver and gold can be found uncombined in the Earth's crust. Most metals are found combined with elements such as oxygen or sulphur in **metal ores.** For example, bauxite is the ore from which aluminium is extracted.

Methods of extraction

The more reactive a metal is, the more difficult it is to extract from its ore.

- The most reactive metals are extracted by **electrolysis** of their molten ores.
- Metals in the middle of the reactivity series are extracted by **heating their ores with carbon or carbon monoxide,**
  e.g. lead oxide + carbon → lead + carbon dioxide.
- The least reactive metals are found **uncombined** in the Earth's crust.

The table below sums this up:

Metal	Method of extraction
Potassium	Electrolysis of molten ore
Sodium	Electrolysis of molten ore
Calcium	Electrolysis of molten ore
Magnesium	Electrolysis of molten ore
Aluminium	Electrolysis of molten ore
Zinc	Heating the ore with carbon or carbon monoxide
Iron	Heating the ore with carbon or carbon monoxide
Tin	Heating the ore with carbon or carbon monoxide
Lead	Heating the ore with carbon or carbon monoxide
Copper	Heating the ore with carbon or carbon monoxide
Silver	Metal occurs uncombined
Gold	Metal occurs uncombined

Top Tip
The more reactive the metal, the more difficult it is to extract from its ore.

Extraction of iron: the blast furnace

The main ore of iron is iron oxide or haematite Fe_2O_3.
Iron is extracted from iron oxide by **reduction**
(removal of oxygen) in the blast furnace.

The solid raw materials added to the blast furnace are:

- iron ore (haematite)
- coke (almost pure carbon)
- limestone (reacts with impurities).

What happens in the blast furnace

1. Hot air enters the blast furnace and reacts with carbon,
   forming carbon dioxide and releasing energy.

   carbon + oxygen → carbon dioxide

   $$C(s) + O_2(g) → CO_2(g)$$

2. At high temperatures the carbon dioxide reacts with more
   carbon to form carbon monoxide.

   carbon dioxide + carbon → carbon monoxide

   $$CO_2(g) + C(s) → 2CO(g)$$

3. The carbon monoxide reduces the iron oxide to iron.

   carbon monoxide + iron oxide → iron + carbon dioxide

   $$3CO(g) + Fe_2O_3(s) → 2Fe(l) + 3CO_2(g)$$

The iron is dense (heavy) and sinks to the bottom, where
it is tapped off.

Removal of impurities

Haematite may contain **impurities**, most
commonly **silicon dioxide** (silica). When
limestone is added it removes the silica by
forming molten **slag**. The slag has a low
density so floats on the top of the molten
iron ore. The slag can be used in road building.

Top Tip
Remember that
the blast furnace is
used to extract
iron from iron
ore.

three solids are added:
iron ore
coke
limestone

VERY HOT

slag (limestone
impurities) is found
on top of the iron

molten or liquid iron
is found at the bottom

hot air blasted
into furnace

The blast furnace

Quick Test

1. Which metals are so unreactive they can be found uncombined?

2. What method of extraction should be used for zinc and lead?

3. What method of extraction should be used for the most reactive metals?

4. What is the name and formula of the main iron ore?

5. What three solid raw materials are added to the blast furnace?

6. Which other reactant is blasted into the furnace?

7. Which gas actually reduces the iron oxide?

8. Why does the iron sink to the bottom?

9. What substance is formed when limestone reacts with silica?

10. What can this substance be used for?

Answers 1. Gold and silver **2.** Heating the metal ore with carbon **3.** Electrolysis **4.** Haematite, Fe_2O_3 **5.** Iron ore, coke, limestone **6.** Hot air **7.** Carbon monoxide **8.** It is more dense. **9.** Slag **10.** Road building

Introducing plastics and synthetic fibres

Natural and synthetic materials

Materials can be classified as **natural or synthetic**.

Natural materials come from **plants and animals** and include:

- wool from sheep
- wood from trees
- leather from animal skins
- cotton from plants
- rubber from plants.

Top Tip
Make sure you know the differences between natural and synthetic materials.

Synthetic materials are made in factories. Many plastics and synthetic fibres are made from **oil**.

- The first process in the refinery is **fractional distillation**. This separates crude oil into fractions according to their boiling points.
- The second process is **cracking**. This breaks the long chain hydrocarbons into the more desirable smaller chain hydrocarbons. The cracking process also produces small unsaturated molecules such as ethene which are used to make plastics.

Advantages and disadvantages of natural and synthetic materials

An advantage that natural materials have over synthetic materials is that natural materials are **biodegradable.** This means that they are attacked by bacteria in the soil and rot away when they are discarded. Synthetic plastics are not usually biodegradable and their durability causes litter and environmental problems when they are discarded.

Synthetic materials have an advantage in that they can be tailored to fit the required property. However, synthetic plastics may burn to produce toxic fumes. If the plastic contains chlorine, toxic hydrogen chloride may be produced. Polyurethane plastics may produce very poisonous hydrogen cyanide if they are burning or smouldering. If plastics are burned in a limited supply of oxygen, carbon monoxide may be formed.

Top Tip
Make sure you know what biodegradable means.

Thermoplastics and thermosetting plastics

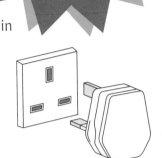

- A **thermoplastic** softens on heating and can be remoulded or reshaped more than once. Examples include **polythene, polystyrene, polyvinylchloride** and other addition polymers.

- A **thermosetting plastic** does not soften on heating and cannot be reshaped. Examples include **bakelite** and **urea-methanal**. Since they are unaffected by heating, thermosetting plastics are used in electric plugs and sockets. Melamine is a thermosetting plastic which is used to make kitchen surfaces as it will not melt when hot pans are placed on it.

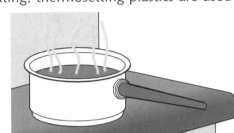

Examples of thermosetting plastics

Uses of plastics

Poly(ethene)
- Poly(ethene) is **cheap** and **strong**.
- It is used for plastic **bags** and **bottles**.

Poly(propene)
- Poly(propene) is **strong** and has a high **elasticity**.
- It is used for **crates** and **ropes**.

Poly(chloroethene), PVC
- PVC is **rigid** and is used for building materials such as **drain pipes**.
- With plasticisers added, it is used for **wellingtons** and **mackintoshes**.
- PVC is also used as insulation around electrical wires and cables.

Poly(styrene)
- Poly(styrene) is **cheap** and can be **moulded** into different shapes.
- It is used for **packaging** and for plastic **casings**.

Quick Test

1. What does synthetic mean?
2. Where do most synthetic materials come from?
3. What does biodegradable mean?
4. Write down two disadvantages of synthetic plastics.
5. What is meant by thermosetting?
6. Name two thermoplastics.
7. Write down a property and use for poly(propene).

Answers 1. Artificial or man-made. **2.** Crude oil. **3.** Attacked by bacteria and rots away naturally. **4.** They cause environmental problems since they are not biodegradable; they may burn to produce toxic fumes. **5.** Does not soften on heating. **6.** Poly(ethene), poly(styrene), poly(propene), etc. **7.** Strong and elastic, used for crates and ropes.

Making plastics

Addition polymerisation

- Many plastics are made from small unsaturated molecules produced by **cracking**.
- The small unsaturated molecules are called **monomers**.
- Plastics are made when the small molecules join together to give much larger molecules called **polymers**. Plastics are examples of polymers.
- Natural and synthetic fibres are also examples of polymers
- The process of monomers joining together to form polymers is called **polymerisation**.
- The making of plastics is an example of polymerisation.

Ethene, a small unsaturated molecule produced in cracking, can be converted into poly(ethene) by polymerisation:

$$
\begin{array}{l}
\overset{\displaystyle H \quad H}{\underset{\displaystyle H \quad H}{C=C}} + \overset{\displaystyle H \quad H}{\underset{\displaystyle H \quad H}{C=C}} + \overset{\displaystyle H \quad H}{\underset{\displaystyle H \quad H}{C=C}} \longrightarrow \ \ \overset{\displaystyle H \ H \ H \ H \ H \ H}{\underset{\displaystyle H \ H \ H \ H \ H \ H}{-C-C-C-C-C-C-}}
\end{array}
$$

- In this reaction ethene is the **monomer** and poly(ethene) is the **polymer**.
- The small unsaturated ethene molecules join together when the carbon-to-carbon double bond opens.
- This type of polymerisation is called **addition polymerisation** because the molecules (monomers) **add** on to each other to form a large molecule (polymer).

Propene is another monomer made by cracking.
Propene can be polymerised to **poly(propene)**:

$$
\begin{array}{l}
\overset{\displaystyle H \ CH_3}{\underset{\displaystyle H \ \ H}{C=C}} + \overset{\displaystyle H \ CH_3}{\underset{\displaystyle H \ \ H}{C=C}} + \overset{\displaystyle H \ CH_3}{\underset{\displaystyle H \ \ H}{C=C}} \longrightarrow \ \ \overset{\displaystyle H \ CH_3 \ H \ CH_3 \ H \ CH_3}{\underset{\displaystyle H \ \ H \ \ H \ \ H \ \ H \ \ H}{-C-C-C-C-C-C-}}
\end{array}
$$

CREDIT

Other monomers which can be made from ethene are:
- chloroethene (vinyl chloride)   which polymerises to   poly(chloroethene) or poly(vinyl chloride)
- tetrafluoroethene   which polymerises to   poly(tetrafluoroethene) or teflon
- phenylethene (styrene)   which polymerises to   poly(phenylethene) or polystyrene
- methyl-2-methylpropenoate   which polymerises to   poly(methyl-2-methylpropenoate) or perspex.

More about addition polymerisation

There are two types of polymerisation; **addition polymerisation** and **condensation polymerisation**. This chapter deals with addition polymerisation only. Polymers made by addition polymerisation are **thermoplastic**. Condensation polymers can be thermoplastic or thermosetting.

The monomer molecules for addition polymerisation must contain a C=C double bond and so are **unsaturated** molecules. These small unsaturated molecules join together when the carbon to carbon double bond opens up and the monomers then **add together** to form the **long chain polymer** molecule.

You should be able to start with the structure of an unsaturated monomer and draw part of the polymer chain, showing three monomer units linked together. Taking a general example, where W, X, Y and Z could be the same or different atoms or groups of atoms:

Top Tip
You should be able to start with the structural formula of a monomer and draw part of the structure of the polymer. You should also be able to work out the structures of the repeating unit and the monomer, given part of the structure of the addition polymer.

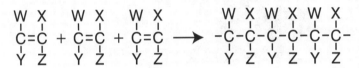

The repeating unit in the above structure is:

```
   W  X
   |  |
 -C--C-
   |  |
   Y  Z
```

Note that the end bonds on the repeating unit and on part of the polymer chain must be left open.

CREDIT

Quick Test

1. Why can ethane not be used as a monomer for addition polymerisation?

2. What will be the name of the polymer formed when the monomer is but-1-ene?

3. The monomer for the formation of poly(tetrafluoroethene) has the formula C_2F_4. Draw the structural formula for the monomer and show the structure of part of the polymer chain.

4. The structure for part of the poly(chloroethene) polymer chain is:

```
   H Cl H Cl H Cl
   | |  | |  | |
 -C-C--C-C--C-C-
   | |  | |  | |
   H H  H H  H H
```

Draw the structure of the monomer and of the repeating unit.

Fertilisers

The need for fertilisers

Increasing world population means that fertilisers are needed to make food production more efficient.

Plant growth involves a series of complicated chemical reactions. Growing plants need **nutrients** and different crops need fertilisers containing the elements **nitrogen (N), phosphorus (P) and potassium (K)** in different proportions.

Which compounds are in fertilisers?

Fertilisers can be natural or artificial. Natural fertilisers include manure and compost.

When plants are growing they are removing nutrients from the soil. Adding fertilisers restores essential elements back into the soil. The main elements needed by plants are nitrogen, phosphorus and potassium. Because plants take up nutrients through their roots, fertilisers need to be soluble in water. Useful synthetic fertilisers include:

a compost bin

- ammonium salts; contain N
- nitrate salts; contain N
- phosphate salts; contain P
- potassium salts; contain K.

Top Tip
Remember NPK; the essential elements in fertilisers.

Page 5 of the Data Booklet shows that ammonium compounds and nitrate compounds are very soluble, so these make good fertilisers. Potassium compounds are also very soluble. The element phosphorus is present in phosphate compounds. Phosphates are not so soluble but ammonium phosphate and potassium phosphate are soluble. Synthetic fertilisers contain mixtures of different compounds containing these essential elements. Different plants need these essential elements in different proportions.

Problems with fertilisers

In areas of high rainfall, some of the artificial fertilisers added to the soil may be washed away by rainwater and end up in lochs and rivers. Here the nitrates encourage bacteria and algae to grow, and this uses up the dissolved oxygen in the water. Fish and other animals therefore cannot get enough oxygen and die.

Urea is only slightly soluble and is a useful fertiliser in areas of high rainfall.

Changing free nitrogen into fixed nitrogen

Nitrogen gas is known as **free nitrogen**. When nitrogen is combined with other elements in compounds it is known as **fixed nitrogen**. Almost 80 per cent of the air is nitrogen gas. Nitrogen is an unreactive gas and will react only under extreme conditions where a lot of energy is supplied by a high voltage spark. Nitrogen and oxygen in the air react together to form nitrogen dioxide, NO_2, during lightning storms.

Certain plants such as peas, beans and clover contain **nitrogen-fixing bacteria** in nodules on their roots.

These bacteria can change nitrogen from the air (free nitrogen) into nitrogen compounds (fixed nitrogen) at ordinary temperatures. The plants can then absorb these nitrogen compounds and convert them into proteins. Peas, beans and clover are good for the soil as they make it more fertile and this is cheaper than using chemical fertilisers.

Top Tip
Remember peas, beans and clover can convert free nitrogen into fixed nitrogen.

The nitrogen cycle

The nitrogen cycle shows how nitrogen is recycled between plants, animals and the air. The decomposition of plant and animal protein is very important in the recycling of nitrogen, and this is a major part of the natural nitrogen cycle. To maintain soil fertility and to feed the increasing world population, artificial fertiliser must be added to the nitrogen cycle. A simplified nitrogen cycle is shown on the right.

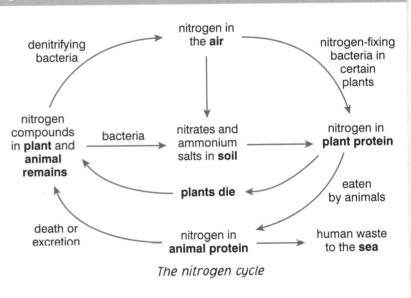

The nitrogen cycle

Quick Test

1. Name a plant that has nitrogen-fixing bacteria in nodules on its roots.
2. Give three reasons why potassium nitrate is useful as a fertiliser.
3. What will happen to algae living in a loch if nitrates are present in the water?
4. Why are fertilisers needed?
5. Which three elements are the main nutrients needed by growing plants?
6. What is the difference between free nitrogen and fixed nitrogen?

Answers 1. Pea, bean or clover. **2.** It contains nitrogen and potassium and is soluble. **3.** The algae grow very well and eventually cause dissolved oxygen in the water to be used up and fish die. **4.** To replace nutrients in the soil removed by growing plants. **5.** Nitrogen, phosphorus and potassium. **6.** Free nitrogen is nitrogen the element; fixed nitrogen is nitrogen combined with other elements in a compound.

Ammonia and the Haber process

Properties of ammonia

- Ammonia has the formula, **NH$_3$**.
- Ammonia is a **colourless gas** with a smell of wet nappies.
- Ammonia gas turns **moist pH paper blue**.
- Ammonia reacts with acids to form **ammonium salts**, for example, it reacts with **nitric acid** to form **ammonium nitrate**, and reacts with **sulphuric acid** to form **ammonium sulphate**.
- Ammonium nitrate and ammonium sulphate are both used as **fertilisers**.
- Ammonia is **very soluble** in water forming an **alkaline solution.**

The solubility of ammonia can be demonstrated in the fountain experiment, as shown in the diagram right.

- The main use for ammonia is making fertilisers.

Top Tip
Try not to get mixed up between the ammonium ion, NH$_4^+$, and ammonia, NH$_3$.

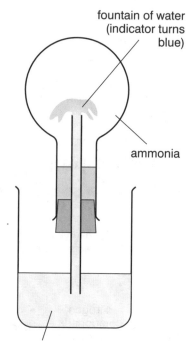

fountain of water (indicator turns blue)

ammonia

The fountain experiment

water containing universal indicator

Making ammonia in the laboratory

Ammonia, NH$_3$, can be made in the lab by heating an ammonium salt with sodium hydroxide solution:

$$NH_4Cl + NaOH \rightarrow NaCl + H_2O + NH_3$$

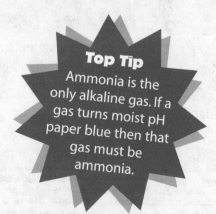

Top Tip
Ammonia is the only alkaline gas. If a gas turns moist pH paper blue then that gas must be ammonia.

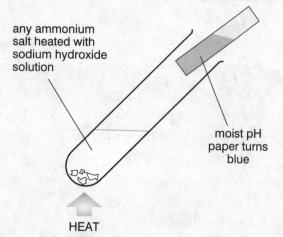

any ammonium salt heated with sodium hydroxide solution

moist pH paper turns blue

HEAT

Heating an ammonium salt and sodium hydroxide solution to produce ammonia

CREDIT

The Haber process

In industry ammonia is made in the **Haber process** by combining nitrogen from the air with hydrogen from methane, CH_4, or from water, H_2O.

This is an example of a **reversible reaction**.

- nitrogen and hydrogen $\rightleftharpoons$ ammonia

Some of the nitrogen and hydrogen react to form ammonia. At the same time, some of the ammonia breaks down into nitrogen and hydrogen.

Not all of the hydrogen and nitrogen are converted to ammonia, thus giving a mixture of hydrogen, nitrogen and ammonia.

When this mixture is **cooled** the ammonia **liquefies** and is removed.

The remaining nitrogen and hydrogen are recycled to reduce costs.

Industrial conditions

- **high pressure** (200 atmospheres) to give a higher yield
- **moderate temperature** (450ºC) – low enough to give a decent yield but not too low otherwise the reaction is too slow
- an iron catalyst.

The catalyst speeds up the rate of reaction and so reduces the cost of producing the ammonia.

These conditions produce a reasonable amount of ammonia fairly quickly.

Nitrogen and hydrogen are mixed

Iron catalyst
Temperature of 450°C
Pressure of 200 atmospheres

Unreacted hydrogen and nitrogen are recycled

Ammonia produced is removed

Top Tip
The Haber process is important. Make sure you know the experimental conditions.

Quick Test

1. What is the test for ammonia gas?

2. Which experiment shows that ammonia is very soluble in water?

3. How can ammonia gas be prepared from ammonium sulphate?

4. Where do the nitrogen and hydrogen reactants come from in the Haber process?

5. Copy and complete the table below

Industrial process	Temperature (°C)	Catalyst	Pressure
Haber			

Nitric acid and the Ostwald process

The Ostwald process

The **Ostwald process** is the name given to the industrial manufacture of nitric acid. The main use for nitric acid is for making fertilisers. The following flow diagram outlines the Ostwald process.

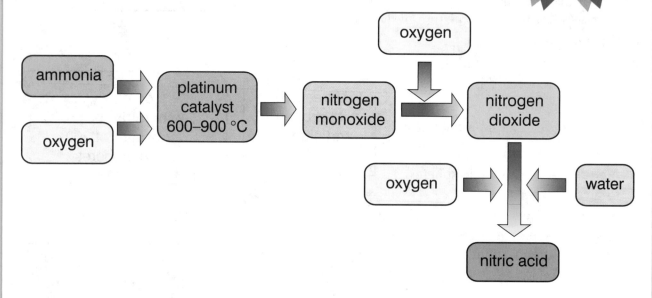

- The first step is the **catalytic oxidation** of some of the ammonia produced in the Haber process. Ammonia and air are passed at atmospheric pressure over a **platinum gauze catalyst** heated to between **600 and 900**°C. This reaction is **exothermic** and so once the catalyst has been heated it will remain at this high temperature. The ammonia is oxidised (reacts with oxygen) to form **nitrogen monoxide, NO**, and water:

  ammonia + oxygen → nitrogen monoxide + water
  $$NH_3 + O_2 \rightarrow NO + H_2O$$

- The **nitrogen monoxide** formed then reacts with more oxygen to form the brown gas, **nitrogen dioxide, NO$_2$**:

  nitrogen monoxide + oxygen → nitrogen dioxide
  $$NO + O_2 \rightarrow NO_2$$

- The nitrogen dioxide is then reacted with water and more oxygen making nitric acid, HNO$_3$:

  nitrogen dioxide + oxygen + water → nitric acid
  $$NO_2 + O_2 + H_2O \rightarrow HNO_3$$

Nitrogen dioxide can also be formed by combining nitrogen and oxygen from the air in the presence of a high energy spark. This requires such a large quantity of electricity that it is not an economic method of preparing nitric acid. The reaction occurs naturally during lightning storms.

The Ostwald process in the laboratory

The catalytic oxidation of ammonia can also be demonstrated in the school laboratory using the apparatus shown below.

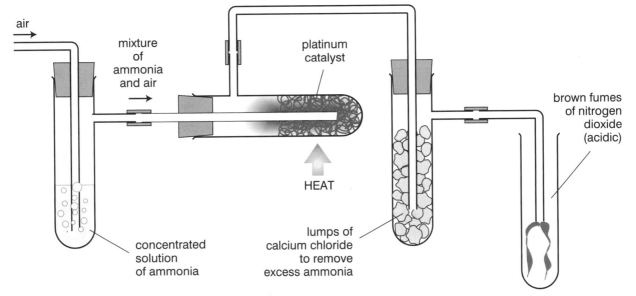

The catalytic oxidation of ammonia

After a time the platinum catalyst can be seen to glow red hot even after it has stopped being heated. This is because it is an exothermic reaction.

The flask collects the brown nitrogen dioxide gas showing that the ammonia has been oxidised. If universal indicator solution is added to the flask, the indicator turns red showing that nitric acid has formed.

Quick Test

1. Write the chemical formula for:
   a) ammonia
   b) nitrogen monoxide
   c) nitrogen dioxide
   d) nitric acid.

2. Copy and complete the table below:

Industrial process	Temperature (°C)	Catalyst
Ostwald		

3. Write chemical equations for the following word equations:
   a) nitrogen + oxygen → nitrogen monoxide
   b) nitrogen monoxide + oxygen → nitrogen dioxide.

4. What colour is nitrogen dioxide gas and what colour does it turn universal indicator?

Answers 1. a) NH_3 **b)** NO **c)** NO_2 **d)** HNO_3 **2**

Industrial process	Temperature (°C)	Catalyst
Ostwald	600–900	Platinum

3. a) $N_2 + O_2 \rightarrow NO$ **b)** $NO + O_2 \rightarrow NO_2$

4. NO_2 is brown and it turns universal indicator red.

Carbohydrates

What are carbohydrates?

The word **carbo-hydr-ate** tells us that the elements **carbon, hydrogen and oxygen** must be present. The experiment shown right demonstrates that carbohydrates contain carbon and water.

The hydrogen and oxygen in carbohydrates are in the ratio 2:1, just as they are in water.

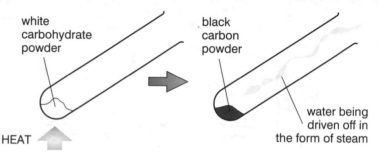

white carbohydrate powder

black carbon powder

water being driven off in the form of steam

HEAT

Heating carbohydrate to produce carbon and water

Names and formulae of carbohydrates

The table right gives the names and formulae of common carbohydrates.

Starch is a **polymer** made from many glucose molecules joined together. When two glucose molecules join together, water is lost:

Name of carbohydrate	Formula
Glucose	$C_6H_{12}O_6$
Fructose	$C_6H_{12}O_6$
Sucrose	$C_{12}H_{22}O_{11}$
Maltose	$C_{12}H_{22}O_{11}$
Starch	$(C_6H_{10}O_5)n$

> **Top Tip**
> Carbohydrates with the formula $C_6H_{12}O_6$ are monosaccharides, e.g. glucose. Carbohydrates with the formula $C_{12}H_{22}O_{11}$ are disaccharides, e.g. sucrose. Starch is a polysaccharide.

$$C_6H_{12}O_6 + C_6H_{12}O_6 \rightarrow C_{12}H_{22}O_{11} + H_2O$$

When many glucose molecules join together, each glucose molecule that joins loses a water molecule. That is why the formula for starch is given as $(C_6H_{10}O_5)n$ (since $C_6H_{12}O_6 - H_2O = C_6H_{10}O_5$). This type of polymerisation, where small molecules join together to make a large molecule and water, is known as **condensation polymerisation**.

Glucose and fructose have the same molecular formula but their structures are different and so glucose and fructose are **isomers**. (Isomers have the same molecular formula but different structures.) Maltose and sucrose are also isomers.

Solubility of carbohydrates

Most carbohydrates such as glucose and sucrose dissolve easily in water to form sweet tasting solutions. These carbohydrates are sometimes called **sugars** and are often added to foods and drinks.

Starch does not taste sweet and does not dissolve in water to form a true solution. If a light beam is shone through a test tube of glucose solution and a test tube containing starch in water, the light beam does not show up in the glucose solution but shows up in the other test tube as the light reflects off the starch molecules. This shows that starch molecules are bigger than glucose molecules.

Tests on carbohydrates

Test for starch

Starch turns **iodine solution** a **blue-black** colour. No other carbohydrate does this so this test can be used to distinguish starch from other carbohydrates.

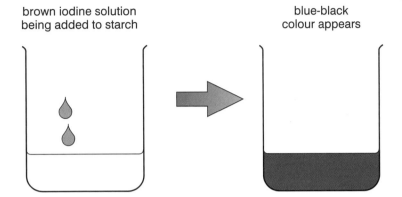

brown iodine solution being added to starch

blue-black colour appears

Starch turns iodine blue-black

Test for glucose

Benedict's (or Fehling's) solution can be used to test for **glucose.** **Blue** Benedict's or Fehling's solution turns **orange-brown** in glucose solution when heated in a hot water bath.

Top Tip
The two tests on this page are important. Make sure you know how to carry them out and what the results mean.

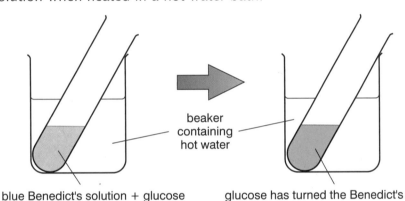

blue Benedict's solution + glucose

beaker containing hot water

glucose has turned the Benedict's solution from blue to orange-brown

Glucose turns Benedict's solution orange-brown

If you were given unlabelled samples of glucose, sucrose and starch, these tests would enable you to tell which was which. Remember, iodine is used to test for starch, Benedict's solution can be used to test for glucose, but there is no test for sucrose

CREDIT

Benedict's solution will react with some other carbohydrates in the same way as it reacts with glucose. These carbohydrates include fructose, maltose and lactose, and carbohydrates which react in this way with Benedict's solution are known as **reducing sugars.**

Quick Test

1. Which elements are present in carbohydrates?

2. Which experiment shows that starch molecules are bigger than glucose molecules?

3. What is the chemical test for starch?

4. How would you distinguish experimentally between glucose and sucrose?

5. Write down the molecular formula for:
   a) glucose
   b) sucrose.

Photosynthesis and respiration

Where do carbohydrates come from?

Carbohydrates are found naturally in plants. The table on the right shows where the common carbohydrates can be found.

Name of carbohydrate	Examples of where found
Glucose	Many plants and fruits
Fructose	Fruits and honey
Sucrose	Sugar cane
Maltose	Barley
Lactose	Milk
Starch	Potatoes

Top Tip
Sucrose is the chemical name for what we call sugar at home.

Photosynthesis

Carbohydrates are made in green plants in a chemical reaction known as **photosynthesis**. Photosynthesis takes place in the leaves of green plants and requires **carbon dioxide** from the air and **water** from the soil.

- The photosynthesis reaction produces **glucose** and **oxygen** gas.
- The photosynthesis reaction **requires energy**. This energy comes from sunlight.

Chlorophyll is the substance in the leaves which absorbs the Sun's energy. The formula equation for photosynthesis is:

$$6CO_2 + 6H_2O \xrightarrow[\text{light energy}]{\text{chlorophyll}} C_6H_{12}O_6 + 6O_2$$

In plants, the glucose formed is converted into larger carbohydrates such as starch by **condensation polymerisation**.

Photosynthesis is one of the most important chemical reactions on Earth as it converts carbon dioxide into oxygen and stores the Sun's energy in the form of carbohydrates which we can eat.

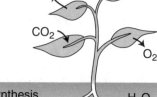

Photosynthesis uses carbon dioxide and water and produces oxygen

Top Tip
Make sure you know the equation for photosynthesis and the conditions needed for the reaction to occur.

Combustion of carbohydrates

When sugars burn they react with oxygen in the air to produce **carbon dioxide** and **water**:

$$C_6H_{12}O_6 + 6O_2 \rightarrow 6CO_2 + 6H_2O$$

This is an **exothermic** reaction because energy is released.

The fact that carbon dioxide and water are produced indicates that carbon and hydrogen must be present in carbohydrates. The carbon in the carbon dioxide must have come from the carbohydrate, and the hydrogen in the water must also have come from the carbohydrate. The oxygen in both the carbon dioxide and in the water may have come from the carbohydrate or from the air or from both.

Energy is released from carbohydrates in our bodies in a much more controlled reaction known as **respiration**. This reaction is the reverse of photosynthesis.

Respiration

Respiration is the process by which animals and plants obtain energy by breaking down carbohydrates using oxygen from the air to give carbon dioxide and water:

$$C_6H_{12}O_6 + 6O_2 \rightarrow 6CO_2 + 6H_2O$$

Glucose is the carbohydrate that reacts with oxygen during respiration. Energy is also produced in the process. The energy produced in respiration is used by animals to move about, produce heat to provide warmth and to send messages through the nerves to and from the brain.

Top Tip

Make sure you know that respiration is the reverse of photosynthesis and that it is how we get energy from our food.

The greenhouse effect

Photosynthesis and respiration are important in maintaining the correct balance of carbon dioxide and oxygen in the atmosphere.

Trees and other plants take in carbon dioxide and convert it to oxygen. The extensive clearing of forests presents dangers to life on Earth as this reduces the amount of carbon dioxide taken from the atmosphere, leading to an overall increase in the amount of carbon dioxide in the atmosphere.

The increase in the amount of carbon dioxide in the atmosphere arising from the clearing of forests is thought to be a significant contributor to global warming caused by changes in the greenhouse effect.

The greenhouse effect is thought to be the reason why the temperature of the Earth is slowly rising. This may lead to changes in climate and if the polar ice caps melt further, a dramatic rise in flooding may occur in some places.

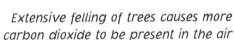

Extensive felling of trees causes more carbon dioxide to be present in the air

Quick Test

1. What is the chemical name for ordinary sugar?

2. Why do plants not photosynthesise during the night?

3. Write the chemical equation for photosynthesis.

4. Give two reasons why photosynthesis is a very important chemical reaction.

5. What type of reaction occurs when small glucose molecules join together to form larger molecules of starch?

6. What word is used to describe chemical reactions that give out energy?

7. What are the products when sucrose burns in air?

8. Write the word equation for respiration.

9. What is the relationship between photosynthesis and respiration?

10. Why is respiration important to all animals?

Answers 1. Sucrose. **2.** Because light is needed. **3.** $6CO_2 + 6H_2O \rightarrow C_6H_{12}O_6 + 6O_2$ **4.** It converts CO_2 into O_2 and stores the Sun's energy as food. **5.** Condensation polymerisation. **6.** Exothermic. **7.** Carbon dioxide and water. **8.** Glucose + oxygen → carbon dioxide + water **9.** One is the reverse of the other. **10.** It is the chemical reaction by which we get energy from carbohydrates.

Breaking down carbohydrates

Structure of carbohydrates

The molecular structure of carbohydrates is very complex. Glucose and fructose both have the molecular formula $C_6H_{12}O_6$. We can represent the structure of these monosaccharides as:

HO—⬡G—OH   and   HO—⬡F—OH

We can represent three glucose molecules joining together to form part of a starch polymer molecule as:

HO—⬡G—OH + HO—⬡G—OH + HO—⬡G—OH 3 glucose molecules

⟶ —O—⬡G—O—⬡G—O—⬡G—O— part of a starch molecule containing 3 glucose molecules joined together
+ H_2O + H_2O + H_2O

Similarly, sucrose (a disaccharide containing one glucose molecule and one fructose molecule joined together) can be represented as:

HO—⬡G—O—⬡F—OH

Digestion of sucrose

When sucrose (ordinary table sugar) is eaten it is **digested** (broken down) in our body into glucose and fructose:

HO—⬡G—O—⬡F—OH ⟶ HO—⬡G—OH + HO—⬡F—OH
+ H_2O

When sucrose is made in plants, one glucose molecule and one fructose molecule join together by **condensation**. It is known as condensation because water molecules are also formed in the reaction.

Breaking down or digesting sucrose takes place in the presence of water. The H from the water molecule joins onto one of the products and the OH from the water molecule joins onto the other product. This is an example of **hydrolysis**. Hydrolysis is the reverse of condensation just as respiration is the reverse of photosynthesis. Hydrolysis is the breaking down of a large molecule into smaller molecules using water.

The chemical reactions taking place in our digestive system are all hydrolysis reactions. They usually take place in the presence of acid or **enzymes** which speed up the hydrolysis reactions. Enzymes are **biological catalysts**. They speed up reactions in living things such as plants and animals.

Top Tip
Remember the reactions, condensation and hydrolysis, and that one is the reverse of the other.

CREDIT

Digestion of starch

When we digest foods such as bread and potatoes that contain large amounts of starch, the starch molecules are broken down into smaller glucose molecules. Starch molecules are too big to travel through the gut wall into the bloodstream. The smaller glucose molecules can pass through the gut wall into the bloodstream where they can be transported around the body.

In the laboratory, starch can be broken down into smaller molecules using either dilute acid or the enzyme amylase. Enzymes such as amylase act as biological catalysts. This means that they speed up chemical reactions in living things such as plants and animals. In our digestive system, enzymes help to break down the complex molecules present in the food we eat into smaller more useful molecules.

Top Tip
Remember that starch is made from many glucose molecules joined together.

The diagram below shows part of the starch polymer molecule breaking into three glucose molecules.

part of a starch molecule containing 3 glucose
molecules joined together

$$-O-\fbox{G}-O-\fbox{G}-O-\fbox{G}-O-$$

$+ H_2O \qquad + H_2O \qquad + H_2O$

$$\longrightarrow \quad HO-\fbox{G}-OH \quad + \quad HO-\fbox{G}-OH \quad + \quad HO-\fbox{G}-OH$$

3 glucose molecules

At the beginning of the above reaction the starch will turn iodine solution a blue-black colour. After some time the starch will no longer react with the iodine solution because its large molecules are breaking down. When this happens the reaction mixture can be tested with Benedict's solution. This will change colour from blue to orange-brown, showing that glucose is made when starch breaks down.

Quick Test

1. Draw a diagram to show two glucose molecules joined together.

2. What type of chemical reaction is occurring when glucose molecules join together to form a disaccharide?

3. What name is given to the chemical reaction in which starch is broken down into glucose?

4. What is an enzyme?

5. Which enzyme is involved in the digestion of starch in our bodies?

Answers 1. $HO-\fbox{G}-O-\fbox{G}-OH$ 2. Condensation. 3. Hydrolysis 4. A biological catalyst. 5. Amylase.

Fermentation

What happens during fermentation?

Fermentation is the chemical reaction in which **glucose** breaks down to form **alcohol** and **carbon dioxide**. The enzyme, **zymase**, produced by **yeast** is the **catalyst** for the reaction.

The equation for the fermentation reaction is:

$$C_6H_{12}O_6 \rightarrow 2C_2H_5OH + 2CO_2$$

Ethanol is the alcohol found in alcoholic drinks.

A fermentation experiment that can be done in the laboratory is shown below:

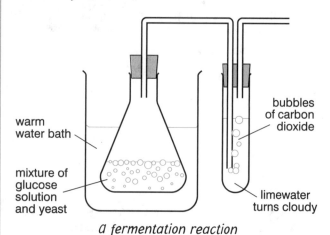

warm water bath

mixture of glucose solution and yeast

bubbles of carbon dioxide

limewater turns cloudy

a fermentation reaction

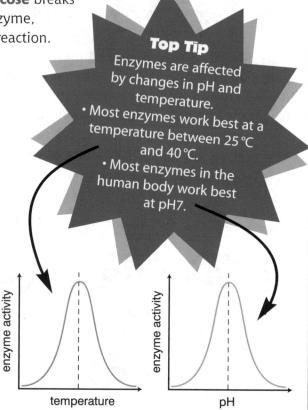

Top Tip
Enzymes are affected by changes in pH and temperature.
• Most enzymes work best at a temperature between 25 °C and 40 °C.
• Most enzymes in the human body work best at pH7.

enzyme activity — temperature

enzyme activity — pH

Alcohol and alcoholic drinks

Alcoholic drinks are made by fermentation. Alcohols are poisonous and the least poisonous alcohol, ethanol, is the one present in alcoholic drinks. Ethanol is the second member of the alkanol family. It has the formula C_2H_5OH and its full structural formula is:

```
    H  H
    |  |
H — C — C — OH
    |  |
    H  H
```

Top Tip
Ethanol is the only alcohol that you need to know. You should be able to write its chemical formula and full structural formula.

Alcoholic drinks have been made for over one thousand years. One of the reasons that there are so many different types of alcoholic drinks is that the glucose for the fermentation can come from different plant sources. Alcoholic drinks can be made from any fruit or vegetable that contains starch or sugars. The table below gives some examples.

Alcoholic drink	Plant used	Approximate % of alcohol
Beer	Barley	4
Wine	Grapes	13
Cider	Apples	5
Sherry	Grapes	20
Whisky	Barley	40

Distillation

When the concentration of alcohol in the fermenting mixture reaches about 10 per cent, the yeasts are killed and the fermentation reaction stops so no more alcohol will form.

Drinks with an alcohol content much above 10 per cent must be **distilled** after fermentation has taken place. These drinks include whisky, brandy, gin and vodka. These alcoholic drinks are usually known as spirits.

Distillation is **evaporation** followed by **condensation**. It is used to separate the alcohol from the aqueous mixture (the alcohol has boiling point 79 °C compared to the boiling point of water which is 100 °C).

During distillation the fermented mixture is warmed gently. When the temperature reaches 79 °C, the ethanol will boil off. The temperature will remain at 79 °C until almost all the ethanol has been removed. If the temperature is allowed to rise to 100 °C then the water will evaporate off.

The apparatus used in the laboratory for distillation is shown below.

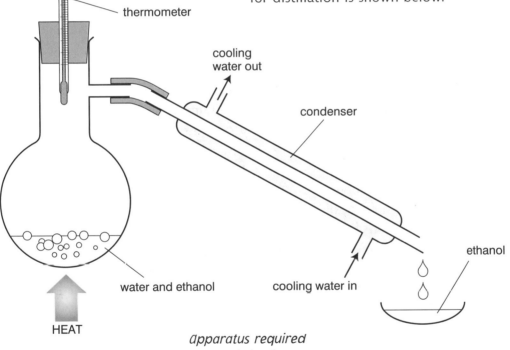

apparatus required for distillation

Quick Test

1. What are the two products formed during the fermentation of glucose?

2. Draw the full structural formula of ethanol.

3. Which technique is used to separate alcohol from water and why does it work?

Calculations 1

Please note that the calculations in this chapter are all at credit level in Standard Grade Chemistry.

Number of protons, electrons and neutrons in ions

This subject has been touched on in an earlier chapter but here it is done in slightly more detail.

Atomic number and mass number

Two important quantities are the **atomic number** and the **mass number**.

The atomic number is the number of protons in an atom. Atoms are **neutral** overall because they have **equal numbers** of **positive protons** in the nucleus and **negative electrons** outside the nucleus. However atoms can lose and gain electrons to form **ions**.

Atoms of **metal elements** tend to lose electrons to form **positive ions**. Atoms of **non-metal elements** tend to gain electrons to form **negative ions**. **The mass number is the sum of the protons and neutrons in an atom.**

So, if we are given the nuclide notation of atoms and ions, we can work out the number of protons, neutrons and electrons as in the examples below.

Top Tip
Make sure you understand the distinction between atoms and ions.

Example

Calculate the number of protons, neutrons and electrons in $^{35}Cl^-$.

- All chlorine atoms and ions have an atomic number of 17, **so the number of protons must be 17.**

- The chloride ion, Cl^-, has a charge of −1 and so has one extra electron compared to a chlorine atom. This means that the number of electrons is one more than the number of protons. So **the number of electrons is 18**.

- The number of neutrons is found by subtracting the atomic number from the mass number. So **the number of neutrons = 35 − 17 = 18**.

Example

Calculate the number of protons, neutrons and electrons in $^{25}Mg^{2+}$

- All magnesium atoms and ions have an atomic number of 12, **so the number of protons must be 12**.

- The magnesium ion, Mg^{2+}, has a charge of +2 so it has two fewer electrons than a magnesium atom. So the number of electrons is two fewer than the number of protons. So **the number of electrons is 10**.

- The number of neutrons is found by subtracting the atomic number from the mass number. So **the number of neutrons = 25 − 12 = 13**.

Relative atomic mass

The **relative atomic mass (RAM)** is used to **compare** the masses of different atoms. The **relative atomic mass** of an element is the average mass of its isotopes compared with an atom of $^{12}_{6}C$.

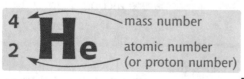

$^{4}_{2}He$ — mass number / atomic number (or proton number)

RAM of helium = 4

$^{24}_{12}Mg$

RAM of magnesium = 24

Most elements exist as a mixture of different isotopes and the relative atomic mass is the average atomic mass. Therefore the relative atomic mass is rarely a whole number. The approximate relative atomic masses are given in the Data Booklet on page 4 and have been rounded to the nearest 0.5

Top Tip
Remember that the relative atomic masses of most elements are given in the Data Booklet. Mass numbers are **not** given in the Data Booklet and these should not be confused with relative atomic masses.

Formula mass

The **formula mass** of any molecule is worked out by adding together the **atomic masses** of all the atoms in the molecule.

- For carbon dioxide, CO_2:

The formula mass of CO_2 is **44**.

$C \; O_2$

$12 + (2 \times 16) = 44$

- For water, H_2O:

The formula mass of H_2O is **18**.

$H_2 \; O$

$(2 \times 1) + 16 = 18$

- For ammonia, NH_3:

The formula mass of NH_3 is **17**.

$N \; H_3$

$14 + (3 \times 1) = 17$

Top Tip
Remember that the relative atomic masses of the more common elements are given in the table on page 4 of the Data Booklet.

The mole

No units are given for formula masses.

Very often in chemistry we need to know actual quantities. To do this we use a value known as **the mole**.

The mole is the gram formula mass of a substance. This is really the same as the formula mass, but with grams as the unit.

Examples

Using the same examples as above:
- 1 mole of carbon dioxide, CO_2, is 44g and 1 mole of water, H_2O, is 18g.
- 2 moles of ammonia will be $2 \times 17g = 34g$ and 0.5 moles of water will be $18 \times 0.5 = 9g$.

Quick Test

1. How many protons, neutrons and electrons are in:
   a. $^{18}O^{2-}$. b. $^{24}Na^+$. c. $^{27}Al^{3+}$. d. $^{79}Br^-$

2. When calculated accurately, why is the relative atomic mass for most elements not a whole number?

3. Calculate the formula mass for:
   a. CO. b. NaOH.
   c. magnesium oxide. d. calcium carbonate

4. Calculate the mass of 2 moles of:
   a. NaOH. b. magnesium oxide.
   c. calcium carbonate. d. sodium chloride.

Answers 1. a. 8 p,10 n. **b.** 11 p, 13 n, 10 e. **c.** 13 p, 14 n, 10 e. **d.** 35 p, 44 n; 36 e. **2.** Because it is an average mass of the different isotopes of the element. **3 a.** 28. **b.** 40. **c.** 40.5. **d.** 100. **4. a.** 80g. **b.** 81g. **c.** 200g. **d.** 117g.

Calculations 2

Moles to mass and mass to moles

To do some calculations it is helpful to know and use the triangle below. This shows the relationship between the number of moles, n, the mass and formula mass, FM, of a substance.

n = number of moles
FM = formula mass

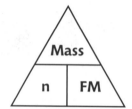

Using this formula
n = Mass/FM
Mass = n × FM
FM = mass/n

Examples

Question 1: Calculate the number of moles in 10 g of sodium hydroxide.

Worked answer: Sodium hydroxide has formula NaOH
Formula Mass, FM, of NaOH = 23 + 16 + 1 = 40
Mass = 10g
n = mass/FM = 10/40 = **0.25 moles**

Question 2: Calculate the mass of 2 moles of calcium chloride.

Worked answer: Calcium chloride has formula $CaCl_2$
Formula mass, FM, of $CaCl_2$ = 40 + (35.5 × 2) = 111
Number of moles, n = 2
Mass = n × FM = 2 × 111 = **222g**

Concentrations of a solution

The concentration of a solution is the **quantity of solute dissolved in a certain volume of solution**.

In chemistry, the concentration is **measured in units of moles per litre (mol/l)**.
For example:
1 mole of sodium hydroxide, NaOH, has a mass of 40.0 g.
So 1 litre of 1mol/l NaOH(aq) will contain 40.0g of sodium hydroxide in 1 litre of solution.

Another triangle, which is useful in calculations involving volume and concentration of solutions, such as dilute acids and alkalis, is shown below.

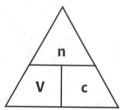

n = number of moles
V = volume of solution (**in litres**)
c = concentration of solution (**in mol/l**)

Using this triangle,
n = V × c
c = n/V

Example

Question 1: Calculate the number of moles in 500 cm³ of a 0.2 mol/l solution.

Worked answer: The volume, V, is 500 cm³ = **0.5 litres**
The concentration, c = **0.2 mol/l**
n = V × c = 0.5 × 0.2 = **0.1 moles**

Calculations using both formulae

Sometimes, in more difficult calculations, it is necessary to use both triangles.

Examples

When preparing solutions of a certain concentration, it is necessary to work out how much of the solute is needed.

Question 1: Calculate the mass of sodium hydroxide required to prepare $250\,cm^3$ of a solution of concentration 0.1 mol/l.

Worked answer: In this calculation we are told:
- the volume, $V = 250\,cm^3$ which is 0.25 litres and
- the concentration, $c = 0.1$ mol/l

We can **calculate the number of moles** using the formula

$n = V \times c = 0.25 \times 0.1 = $ **0.025 moles**

The formula of sodium hydroxide is NaOH, so the formula mass can be calculated as:
$FM = 23 + 16 + 1 = 40$.

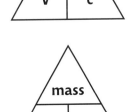

To calculate the **mass** we have to use the other formula

$Mass = n \times FM = 0.025 \times 40 = $ **1.0 g**

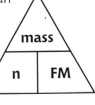

So the mass of sodium hydroxide required is 1.0 g

Question 2. What will be the concentration of $500\,cm^3$ of solution containing 50.0 g of sodium hydroxide?

Worked answer: In this calculation we are told:
- the volume, $V = 500\,cm^3$ which is 0.50 litres and
- the mass which is 50.0 g of NaOH.

We can work out the formula mass of NaOH to be 40.0, so we can work out the number of moles of NaOH using this triangle.

The number of moles, $n = mass/FM = 50.0/40.0 = $ **1.25 moles**

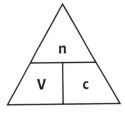

The concentration can now be calculated using the other triangle.

$c = n / V = 1.25 / 0.5 = $ **2.5 mol/l**

So the concentration of the sodium hydroxide solution will be 2.5 mol/l.

Calculations 3 *CREDIT*

Calculations from results of titrations

This type of calculation is used to find out the concentration of an acid or alkali after having done a **titration**.

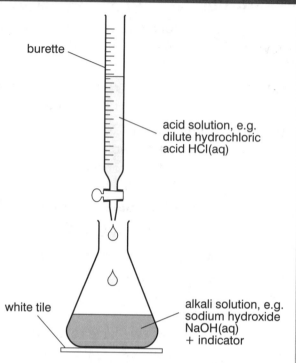

burette

A titration experiment involves using a burette, which is used to dispense accurately measured volumes of a liquid - usually, the acid solution. The acid from the burette flows into a conical flask which would normally contain a known volume of the alkali plus a few drops of indicator. The indicator changes colour when the neutralisation reaction is complete.

acid solution, e.g. dilute hydrochloric acid HCl(aq)

If the volumes of both the acid and alkali are known and the concentration of one of them also known, then the concentration of the other can be calculated.

To do this we can use the formula:

$$P_{acid} \times V_{acid} \times c_{acid} = P_{alkali} \times V_{alkali} \times c_{alkali}$$

white tile

alkali solution, e.g. sodium hydroxide NaOH(aq) + indicator

Where: P = power
V = volume
c = concentration

- **The power of an acid is the number of H^+ ions in the formula**
  (e.g. hydrochloric acid HCl has a power of 1; sulphuric acid H_2SO_4 has a power of 2).

- **The power of an alkali is the number of OH^- ions in the formula**
  (e.g. sodium hydroxide NaOH has a power of 1; calcium hydroxide $Ca(OH)_2$ has a power of 2).

Example

Question.   In a titration experiment, 16.2 cm³ of 0.1 mol/l sulphuric acid was needed to neutralise 25.0 cm³ of sodium hydroxide solution. Calculate the concentration of the sodium hydroxide solution.

Worked Answer: H_2SO_4 has power = 2, so P_{acid} = 2
The volume of the acid is 16.2 cm³, so V_{acid} = 16.2
The acid has concentration 0.1 mol/l so c_{acid} = 0.1

NaOH has power = 1, so P_{alkali} = 1
The volume of the alkali is 25.0 cm³, so V_{alkali} = 25.0
The unknown is c_{alkali}

$$P_{acid} \times V_{acid} \times c_{acid} = P_{alkali} \times V_{alkali} \times c_{alkali}$$

$$2 \times 16.2 \times 0.1 = 1 \times 25.0 \times c_{alkali}$$

$$\text{So } c_{alkali} = \frac{2 \times 16.2 \times 0.1}{1 \times 25.0}$$

$$= \textbf{0.13 mol/l}$$

So the concentration of the sodium hydroxide was 0.13 mol/l

- Calculations from the results of titrations are used to find out the concentration of an acid or alkali.
- If the volumes of both the acid and the alkali are known, and the concentration of one of them known, then the concentration of the other can be calculated from the results of the titration experiment.

Quick Test

1. Calculate the number of moles in
   a. 9 g of H_2O
   b. 11.7 g of NaCl
   c. 1.0 g of $CaCO_3$

2. Calculate the number of moles in
   a. 12.7 g of iodine
   b. 8.8 g of carbon dioxide
   c. 2.6 g of calcium fluoride

3. Calculate the mass of
   a. 0.1 moles of CO
   b. 3 moles of $CuSO_4$
   c. 0.02 moles of calcium chloride
   d. 3.1 moles of carbon

4. Calculate the number of moles in
   a. 100 cm³ of 0.1 mol/l solution
   b. 50 cm³ of 0.25 mol/l solution
   c. 250 cm³ of 2.0 mol/l solution
   d. 500 cm³ of 0.01 mol/l solution

5. Calculate the mass of sodium hydroxide required to prepare
   a. 100 cm³ of 2 mol/l solution
   b. 2 litres of 0.012 mol/l solution
   c. 50 cm³ of 0.2 mol/l solution
   d. 250 cm³ of 0.3 mol/l solution

6. Calculate the mass of sodium nitrate required to prepare
   a. 100 cm³ of 0.2 mol/l solution
   b. 2 litres of 0.01 mol/l solution
   c. 50 cm³ of 0.2 mol/l solution
   d. 250 cm³ of 0.1 mol/l solution

7. What will be the concentration of 200 cm³ of a solution containing
   a. 2.4 g of LiOH?
   b. 0.8 g of NaOH?
   c. 1 g of calcium hydroxide?
   d. 0.17 g of silver(I) nitrate?

8. In a titration, 25.0 cm³ of 0.10 mol/l NaOH was exactly neutralised by 22.2 cm³ of hydrochloric acid. Calculate the concentration of the hydrochloric acid.

9. What volume of 0.2 mol/l sulphuric acid will exactly neutralise 25.0 cm³ of 0.5 mol/l potassium hydroxide?

Answers 1. a. 0.5 moles **b.** 0.2 moles **c.** 0.01 moles **2. a.** 0.05 moles **b.** 0.2 moles **c.** 0.033 moles **3. a.** 2.8 g **b.** 478.5 g **c.** 2.22 g **d.** 37.2 g **4. a.** 0.01 moles **b.** 0.0125 moles **c.** 0.5 moles **d.** 0.005 moles **5. a.** 8.0 g **b.** 0.96 g **c.** 0.40 g **d.** 3.0 g **6. a.** 1.7 g **b.** 1.7 g **c.** 0.85 g **d.** 2.125 g **7. a.** 0.5 mol/l **b.** 0.1 mol/l **c.** 0.068 mol/l **d.** 0.005 mol/l **8.** 0.11 mol/l **9.** 31.25 cm³

Calculations 4 CREDIT

Empirical formula calculations

The **empirical formula** of a compound gives **the simplest ratio of the atoms present** in that compound.

For example, ethane has molecular formula C_2H_6.

- This means that in one molecule of ethane there are two carbon atoms and six hydrogen atoms.
- The ratio of carbon atoms to hydrogen atoms is 2:6 and so the simplest ratio is 1:3
- So, the empirical formula will be CH_3.

The empirical formula is obtained by calculating the simplest ratio using the results of an experiment.

Example

The empirical formula of copper oxide can be determined by reducing copper oxide to copper as shown in the diagram.

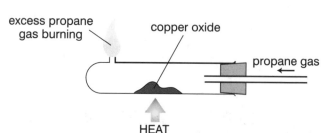

excess propane gas burning

copper oxide

propane gas

HEAT

A set of experimental results is:

- Mass of test tube   = 49.43 g
- Mass of test tube + copper oxide at start   = 51.42 g
- So mass of copper oxide = 51.42 – 49.43   = 1.99 g
- Mass of test tube + copper at the end   = 51.02 g
- So mass of copper = 51.02 – 49.43   = 1.59 g
- Mass of oxygen in the copper oxide = 1.99 – 159 = 0.40 g

To get the empirical formula we must first calculate the number of moles of each element using the formula, n = mass/FM

$n_{for\ Cu}$ = 1.59/63.5 = 0.025 moles of Cu $n_{for\ O}$ = 0.40/16 = 0.025 moles of O

Since there are equal numbers of moles of Cu and O, then the empirical formula is **CuO**.

Example

This is a more difficult example.

Question.   Calculate the empirical formula of a compound containing 1.84 g of sodium, 1.28 g of sulphur and 2.56 g of oxygen.

Worked Answer: Sometimes it is easier to set out empirical formula calculations in a table as shown below.

Element	Na	S	O
Mass/g	1.84	1.28	2.56
Number of moles (= mass/FM)	1.84/23	1.28/32	2.56/16
	0.08	0.04	0.16
Divide by the smallest number (to get the whole number ratio)	0.08/0.04 = 2	0.04/0.04 = 1	0.16/0.16 = 4

So the empirical formula is $Na_2S_1O_4$ or simply, Na_2SO_4

Percentage composition

Percentage mass of an element in a compound $= \dfrac{\text{atomic mass} \times \text{no. of atoms}}{\text{formula mass}} \times 100\%$

Example

Ammonium nitrate, NH_4NO_3, is used as a fertiliser.
Find the percentage composition of nitrogen in this compound.

- RAM of N = 14
- RAM of H = 1
- RAM of O = 16

The formula mass of NH_4NO_3 is:

- $14 + (4 \times 1) + 14 + (3 \times 16) = 80$

Percentage of nitrogen $= \dfrac{14 \times 2}{80} \times 100\% = 35\%$

- **Ammonium nitrate is 35% nitrogen**

Key Fact

1 The empirical formula of a compound gives the simplest ratio of the atoms present in that compound.
2 Obtain the empirical formula by calculating the simplest ratio using the results of an experiment.

Quick Test

1. Calculate the empirical formula of the compound which contains 1.08 g of aluminium combined to 0.96 g of oxygen.

2. Calculate the empirical formula of the hydrocarbon containing 0.60 g of carbon and 0.15 g of hydrogen.

3. Calculate the empirical formula for the compound containing 6.21 g of lead combined to 7.62 g of iodine.

4. 6.40 g of iron oxide was found to contain 4.48 g of iron. Calculate the empirical formula for the iron oxide.

5. Calculate the percentage of hydrogen in ammonium nitrate, NH_4NO_3.

6. Calculate the percentage of oxygen in ammonium nitrate, NH_4NO_3.

7. Calculate the percentage of hydrogen in water.

8. Calculate the percentage of sulphur in sulphur dioxide.

Answers 1. Al_2O_3 **2.** CH_3 **3.** PbI_2 **4.** Fe_2O_3 **5.** 5% **6.** 60% **7.** 11.1% **8.** 50%

Calculations from a balanced equation

Key Facts
- To calculate how much of a product is made, the equation for a reaction can be used.
- The equation for a reaction can also be used to calculate how much reactant should be used.

Calculating the mass of products

The **equation** for a reaction can be used to **calculate** how much product is made.

Example

What **mass** of water is produced when **8 g** of **hydrogen** is burned in **oxygen**?

RAM of H = 1

RAM of O = 16

Write out the balanced equation:

$$2H_2(g) + O_2(g) \rightarrow 2H_2O(g)$$

Work out the formula mass:

$$2 \times (1 \times 2) + (16 \times 2) \rightarrow 2 \times (18)$$
$$4 + 32 \rightarrow 36$$

So 4 g of hydrogen will make 36 g of water.

What does 1 g of hydrogen make?

1 g of hydrogen makes $\frac{36}{4}$ = 9 g of water

Multiply by the number of grams:

8 g of hydrogen makes 9×8 = 72 g of water.

This calculation shows that if 8 g of hydrogen is burned in oxygen, 72 g of H_2O is made.

Calculating the mass of reactants

The equation for a reaction can also be used to calculate how much reactant should be used.

Example

What **mass** of magnesium should be used to produce **60 g** of magnesium oxide?

RAM of Mg = 24.5

RAM of O = 16

Write out the balanced equation:

$2Mg(s) + O_2(g) \rightarrow 2MgO(s)$

Work out the formula masses and the reacting quantities:

$(2 \times 24.5) + (2 \times 16) \rightarrow 2MgO$

$49 + 32 \rightarrow 81$

81 g of MgO is made from 49 g of Mg.

How much is needed to make 1 g of MgO?

1 g of MgO is made from $\frac{49}{81}$ g of Mg.

Multiply by the number of grams:

60 g of MgO is made from $\frac{49}{81} \times 60$ g $= 36.3$ g

This calculation shows that to produce 60 g of MgO, 36.3 g of Mg should be used.

Quick Test

1. Calculate the mass of water produced when 0.2 g of hydrogen is burned in air.

2. Calculate the mass of hydrogen needed to produce 0.72 g of water.

3. Calculate the mass of calcium oxide formed when 10 g of calcium carbonate decomposes as in the equation, $CaCO_3 \rightarrow CaO + CO_2$

4. The equation for calcium burning in oxygen is $2Ca(s) + O_2(g) \rightarrow 2CaO(s)$
   Calculate the mass of calcium required to produce 1.12 g of calcium oxide.

Answers 1. 1.8 g **2.** 0.08 g **3.** 5.6 g **4.** 0.80 g

Types of reactions

The following is a summary of different types of reactions you need to know in Standard Grade Chemistry.

- **Addition** – When a molecule such as hydrogen or bromine 'adds on' across the double C=C bond in an unsaturated hydrocarbon such as ethene.

- **Addition Polymerisation** – When unsaturated monomers join to form a larger molecule known as a polymer by opening their carbon-to-carbon double bonds and linking together.

- **Combustion** – the chemical name for burning. When something burns it reacts with oxygen. If there is not enough oxygen present, **incomplete combustion** may take place.

- **Condensation** – when small molecules join together to make a larger molecule and water is produced.

- **Condensation Polymerisation** – When small molecules, known as monomers, join together to form a larger molecule known as a polymer and water is produced where these molecules link.

- **Corrosion** – When the surface of a metal changes from an element to a compound. In the case of iron this is called rusting. Since a metal loses electrons when it corrodes, corrosion is an example of an oxidation reaction.

- **Cracking** – Breaking large hydrocarbon molecules into smaller molecules which are more useful and more in demand. At least one of the products is unsaturated (contains a double C=C bond).

- **Displacement** – When a metal higher in the electrochemical series takes the place of a metal lower in the electrochemical series. A displacement reaction involves both oxidation and reduction and so is also a redox reaction.

- **Exothermic** – A reaction in which energy is given out. Many different reactions are exothermic, but you should make a point of remembering that combustion and neutralisation are both exothermic.

- **Endothermic** – A reaction in which energy is taken in. Two good examples of endothermic reactions are nitrogen and oxygen reacting together during lightning storms and photosynthesis.

- **Fermentation** – This is the reaction in which sugars are converted into alcohol and carbon dioxide using enzymes found in yeast.

- **Hydrolysis** – Breaking down a large molecule into smaller molecules in the presence of water. Hydrolysis is the reverse of condensation.

- **Neutralisation** – when an acid and a base cancel each other out. Examples of bases include alkalis, metal oxides and metal carbonates.

- **Oxidation** – originally the name given when something reacts with oxygen. Usually an oxidation reaction involves the loss of electrons. (Try to remember **OIL** meaning **O**xidation **I**s **L**oss (of electrons)).

- **Photosynthesis** – The reaction in which plants convert carbon dioxide and water into glucose and water. Light energy and chlorophyll are also required.

- **Precipitation** – When two liquids are mixed together and an insoluble solid is made. The solid formed is called the precipitate.

- **Polymerisation** – The reaction in which small molecules called monomers join together to form larger molecules known as polymers. The two types of polymerisation are **addition polymerisation** and **condensation polymerisation**.

- **Redox** – A term to describe both oxidation and reduction which both take place at the same time.

- **Reduction** – a gain of electrons. (Try to remember **RIG** meaning **R**eduction **I**s **G**ain (of electrons)). Oxidation and reduction are the reverse of each other.

- **Respiration** – The reaction in which oxygen combines with glucose in living organisms to release energy. Carbon dioxide and water are also formed. Respiration is the reverse of photosynthesis.

Index